BARRON'S

Regents Exams and Answers

Algebra I
Fourth Edition

Gary M. Rubinstein, M.S.

Eisa

Published by Kaplan North America, LLC, dba Barron's Educational Series
1515 W Cypress Creek Road
Fort Lauderdale, FL 33309
www.barronseduc.com

ISBN: 978-1-5062-9129-1

10 9 8 7 6 5 4 3 2 1

Kaplan North America, LLC, dba Barron's Educational Series print books are available at special quantity discounts to use for sales promotions, employee premiums, or educational purposes. For more information or to purchase books, please call the Simon & Schuster special sales department at 866-506-1949.

Contents

Regents Exams, Answers, and
Self-Analysis Charts   153

How to Use This Book

This book is designed to help you get the most out of your review for the latest Regents exam in Algebra I. Use this book to improve your understanding of the Algebra I topics, practice answering actual Regents questions, and learn tips and strategies for improving your grade.

About the Exam and Test-Taking Tips

Start with the section "Overview of the Regents Algebra I Exam", which describes each part of the test and how the test is scored. Next, review the section "Test-Taking Tips and Strategies", which covers key advice to keep in mind when working through each type of test question. Remember: no single problem-solving strategy works for all problems—you should have a toolbox of strategies to pick from as you're facing unfamiliar or difficult problems on the test.

Key Algebra I Topics: A Quick Review and Practice

The next section of this book provides you with key Algebra I facts, useful skills, and practice problems with solutions. It offers a quick and easy way to refresh the skills you learned in class. Attempt all practice questions within each topic before reviewing the solutions. As you read through the solutions, make note of the concepts you know really well and the ones that you had difficulty with and may need to review further.

Important Terms to Know

Following the practice section is a glossary that contains key terms that have appeared most frequently on recently administered Algebra I exams. All terms and their definitions are conveniently organized for a quick reference.

Regents Exams and Answers

Once you've reviewed all of the frequently tested topics and key terms, you're ready to take the Regents exams! The next section of this book contains actual Algebra I Regents exams that were administered in June 2019, August 2019, and January 2020.* By answering the questions on these exams, you will be further able to identify your strengths and concentrate on the areas where you may need more study.

Each exam is followed by detailed answers and explanations. Remember, the answer explanations in this book are more than just simple solutions to the problems—they contain facts and explanations that are crucial to success in the Algebra I course and on the Regents exam. Careful review of these answers will increase your chances of doing well.

After you review the answer explanations for each exam, be sure to consult the Self-Analysis Chart at the end of the test. This chart will further help you identify areas that need improvement and direct your study efforts where needed. In addition, the chart classifies the questions on each exam into an organized set of topic groups. This feature will also help you to locate other questions on the same topic in the other Algebra I exams.

Sample 2024 Regents Algebra I Exam

In addition to the three recently administered Regents exams, this book also contains a sample 2024 Regents Algebra I exam. Beginning in June 2024, a newer version of the Regents Algebra I exam will be offered, with updated topics and question types. The Sample 2024 Exam in this book mirrors the format, content tested, and level of difficulty of the upcoming 2024 Regents Algebra I exam and provides you with a clear idea of what to expect on test day. This exam is followed by detailed answers and solutions that explain exactly how to approach each problem.

For Students

Whether you are using this book at the start of the school year or in the weeks leading up to the exam, this book will provide you with the practice you need to maximize your score. Remember that on test day, you will be expected not only to solve algebra problems and answer algebra-related questions but also to *explain your reasoning*. This book will prepare you for both of those tasks by introducing you to the concepts covered

---

*A new version of the Regents Algebra I exam will be administered starting in June 2024. Although some topics from the 2019 and 2020 exams will no longer be tested (and such topics have been noted, where appropriate, within those exams), these tests still provide ample practice with common Algebra I topics that are likely to be tested on the 2024 exam. Students who achieve a high score on these 2019 and 2020 exams will be well prepared for the 2024 exam.

throughout the Algebra I Regents course and the different types of problems (and their varying levels of difficulty) you will encounter on test day.

Becoming familiar with the specific types of questions on the Algebra I Regents exam is crucial to performing well on this test. There are questions in which the math may be fairly easy but the way in which the question is asked makes the question seem much more difficult. For example, the question "Find all zeros of the function $f(x) = 2x + 6$" is a fancy way of asking the much simpler sounding "Solve for x if $2x + 6 = 0$." *Knowing exactly what the questions are asking is a big part of being successful on this test.*

For Teachers

This book is fully up-to-date with the latest Algebra I Regents exam and all upcoming test changes. You can use this book as a tool to help structure an Algebra I course that will culminate with the Regents exam. The topics in the book are arranged by priority, so the sections in the beginning of the book are the ones from which more of the questions on the test are drawn. There are 13 sections, dedicated to all topics for Algebra I, each with practice exercises and solutions. You can use this book as a resource in the classroom, or you can assign practice questions as homework or test material.

Overview of the Regents Algebra I Exam

What Is the Format of the Regents Algebra I Exam?

The Regents Algebra I exam is a three-hour exam that is broken down into 4 parts, as follows.

Part	Question/Item Type	Number of Questions	Total Points
I	Multiple-Choice	24	48 points
II	Short Free-Response	8	16 points
III	Long Free-Response	4	16 points
IV	Long Free-Response	1	6 points

Part I consists of 24 multiple-choice questions. Each question has four possible choices, and you must select the choice that either best completes the statement or answers the question. Each correct answer is worth 2 points, for a total of 48 points for that section of the test.

Part II contains 8 questions, each of which is worth 2 points when answered correctly. For each question, you must indicate all necessary steps, including any appropriate substitutions, diagrams, graphs, and tables, in addition to providing the correct numerical answer. You must show your work; a correct numerical answer without work shown will receive only 1 point.

Part III consists of 4 questions, each of which may involve multiple parts and more detailed calculations. As with the questions in Part II, it is important to show all your work in your answers to the Part III questions, including appropriate formula substitutions, illustrations, graphs, and tables, in addition to the correct numerical answer. For a fully correct and detailed response, you will earn 4 points per question. However, if you do not show your work, the most you can receive for a correct numerical answer is 1 point per question.

Finally, Part IV contains only 1 question, which is worth 6 points. This question involves multiple parts and requires a lot of thought and calculation. To earn most or all of the 6 points, you must indicate all the necessary steps to arrive at the correct answer, so be sure to show all your work, as in Parts II and III. A correct numerical answer without work shown is worth only 1 point.

How Is the Exam Scored?

The Algebra I Regents exam is scored out of a possible 86 points. Unlike most tests given throughout the year by your teacher, the score is not then turned into a percent out of 86. Instead, each test has a conversion sheet, with raw and scaled scores, that varies from year to year.

For example, a raw score of 30 points may become a scaled score of 65, 57 points may become a 75, and 73 points may become an 85. This means that for this hypothetical exam, a student who got 30 out of 86, which is just 35% of the possible points, would get a 65 on this exam. Earning 57 out of 86 is 66% but on this test would be scaled to a 75. Receiving 73 out of 86 points, however, is actually 85% and would become an 85 on this hypothetical exam. Traditionally, there has been a curve on the exam for lower scores, though the scaling is not released until after the exam.

Test-Taking Tips and Strategies

Knowing the material is only part of the battle in acing the Algebra I Regents exam. Things like improper management of time, careless errors, and struggling with the calculator can cost valuable points. This section contains some test-taking strategies to help you perform your best on test day.

TIP 1

Manage Your Time Wisely

Suggestions

- *Don't rush.* The Algebra I Regents exam is three hours long. While you are officially allowed to leave after 90 minutes, you really should stay until the end of the exam. Just as it wouldn't be wise to come to the test an hour late, it is almost as bad to leave the test an hour early.

- *Do the test twice.* If you have enough time left after completing the test, the best way to protect against careless errors is to work through the test a second time and compare the answers you got the first time with the answers you got the second time. For any answers that don't agree, do a "tiebreaker" third attempt. Redoing the test and comparing answers is much more effective than simply looking over your work. Students tend to miss careless errors when looking over their work. By redoing the questions, you are less likely to make the same mistake.

- *Bring a watch.* What will happen if the clock is broken? Without knowing how much time is left, you might rush and make careless errors. Yes, the proctor will probably write the time elapsed on the board and update it every so often, but it's better to be safe than sorry.

The TI-84 graphing calculator has a built-in clock. Press the [MODE] to see it. If the time is not right, go to SET CLOCK and set it correctly. The TI-Nspire does not have a built-in clock.

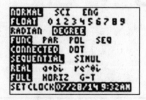

TIP 2

Know How to Get Partial Credit

Suggestions

- *Know the structure of the exam.* The Algebra I Regents exam has 37 questions. The first 24 of those questions are multiple-choice worth 2 points each. There is no partial credit if you make a mistake on one of those questions. Even the smallest careless error, like missing a negative sign, will result in no credit for that question. Parts II, III, and IV are free-response questions with no multiple-choice. Besides giving a numerical answer, you may be asked to explain your reasoning. Part II has 8 free-response questions worth 2 points each. The smallest careless error will cause you to lose 1 point, which is half the value of the question. Part III has 4 free-response questions worth 4 points each. These questions generally have multiple parts. Part IV has 1 free-response question worth 6 points. This question will have multiple parts.

- *Explain your reasoning.* When a free-response question asks you to "Justify your answer," "Explain your answer," or "Explain how you determined your answer," the grader is expecting a few clearly written sentences. For these, you don't want to write too little since the grader needs to see that you understand why you did the different steps you did to solve the equation. You also don't want to write too much because if anything you write is not accurate, points can be deducted.

Here is an example followed by two solutions. The first would not get full credit, but the second would.

Example

Use algebra to solve for x in the equation $\frac{2}{3}x + 1 = 11$. Justify your steps.

Solution 1 (partial credit):

$\frac{2}{3}x + 1 = 11$ $\phantom{\frac{2}{3}x}-1 = -1$ $\frac{2}{3}x = 10$ $\phantom{\frac{2}{3}}x = 15$	I used algebra to get the x by itself. The answer was $x = 15$.

Solution 2 (full credit):

$\frac{2}{3}x + 1 = 11$ $\phantom{\frac{2}{3}x}-1 = -1$ $\frac{2}{3}x = 10$ $\frac{3}{2} \cdot \frac{2}{3}x = \frac{3}{2} \cdot 10$ $1x = 15$ $x = 15$	I used the subtraction property of equality to eliminate the $+1$ from the left-hand side. Then to make it so the x had a 1 in front of it, I used the multiplication property of equality and multiplied both sides of the equation by the reciprocal of $\frac{2}{3}$, which is $\frac{3}{2}$. Then since $1 \cdot x = x$, the left-hand side of the equation just became x and the right-hand side became 15.

- *Computational errors vs. conceptual errors*

  In the Part III and Part IV questions, the graders are instructed to take off 1 point for a "computational error" but half credit for a "conceptual error." This is the difference between these two types of errors.

  If a 4-point question was $x - 1 = 2$ and a student did it like this,

$$x - 1 = 2$$
$$+1 = +1$$
$$x = 4$$

the student would lose 1 point out of 4 because there was one computational error since $2 + 1 = 3$ and not 4.

Had the student done it like this,

$$x - 1 = 2$$
$$-1 = -1$$
$$x = 1$$

the student would lose half credit, or 2 points, since this error was conceptual. The student thought that to eliminate the −1, he should subtract 1 from both sides of the equation.

Either error might just be careless, but the conceptual error is the one that gets the harsher deduction.

TIP 3

Know Your Calculator

Suggestions

- *Which calculator should you use?* The two calculators used for this book are the TI-84 and the TI-Nspire. Both are very powerful. The TI-84 is somewhat easier to use for the functions needed for this test. The TI-Nspire has more features for courses in the future. The choice is up to you. This author prefers the TI-84 for the Algebra I Regents. Graphing calculators come with manuals that are as thick as the book you are holding. There are also plenty of video tutorials online for learning how to use advanced features of the calculator. To become an expert user, watch the online tutorials or read the manual.

- *Clearing the memory.* You may be asked at the beginning of the test to clear the memory of your calculator. When practicing for the test, you should clear the memory too so you are practicing under test-taking conditions.

  This is how you clear the memory.

For the TI-84:

Press [2ND] and then [+] to get to the MEMORY menu. Then press [7] for Reset.

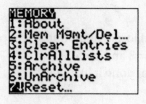

Use the arrows to go to [ALL] for All Memory. Then press [1].

Press [2] for Reset.

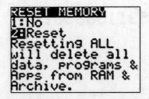

The calculator will be reset as if in brand-new condition. The one setting that you may need to change is to turn the diagnostics on if you need to calculate the correlation coefficient.

For the TI-Nspire:

The TI-Nspire must be set to Press-To-Test mode when taking the Algebra I Regents. Turn the calculator off by pressing [ctrl] and [home]. Press and hold [esc] and then press [home].

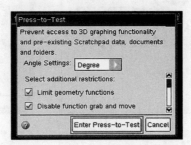

While in Press-to-Test mode, certain features will be deactivated. A small green light will blink on the calculator so a proctor can verify the calculator is in Press-to-Test mode.

To exit Press-to-Test mode, use a USB cable to connect the calculator to another TI-Nspire. Then from the home screen on the calculator in Press-to-Test mode, press [doc], [9] and select Exit Press-to-Test.

■　*Use parentheses*

The calculator always uses the order of operations where multiplication and division happen before addition and subtraction. Sometimes, though, you may want the calculator to do the operations in a different order.

Suppose at the end of a quadratic equation, you have to round $x = \dfrac{-1 + \sqrt{5}}{2}$ to the nearest hundredth. If you enter $(-)$ (1) $(+)$ (2ND) (x^2) (5) $(/)$ (2), it displays

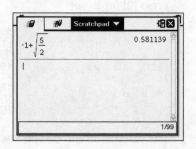

which is not the correct answer.

One reason is that for the TI-84 there needs to be a closing parenthesis (or on the TI-Nspire, press [right arrow] to move out from under the radical sign) after the 5 in the square root symbol. Without it, it calculated $-1 + \sqrt{\dfrac{5}{2}}$. More needs to be done, though, since

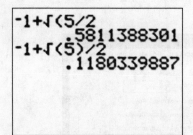
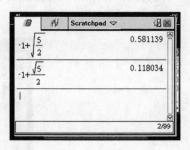

still is not correct. This is the solution to $-1 + \dfrac{\sqrt{5}}{2}$.

To get this correct, there also needs to be parentheses around the entire numerator, $-1 + \sqrt{5}$.

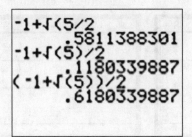
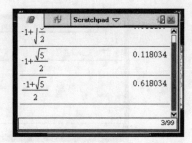

This is the correct answer.

On the TI-Nspire, fractions like this can also be done with [templates].

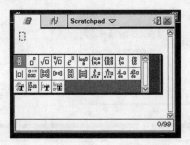
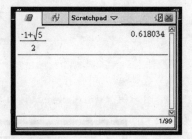

- *Using the ANS feature*

The last number calculated with the calculator is stored in something called the ANS variable. This ANS variable will appear if you start an expression with a +, −, ×, or ÷. When an answer has a lot of digits in it, this saves time and is also more accurate.

If for some step in a problem you need to calculate the decimal equivalent of $\frac{1}{7}$, it will look like this on the TI-84:

For the TI-Nspire, if you try the same thing, it leaves the answer as $\frac{1}{7}$. To get the decimal approximation, press [ctrl] and [enter] instead of just [enter].

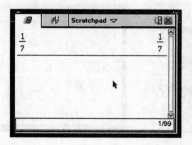 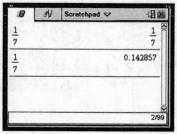

Now if you want to multiply this by 3, just press [×], and the calculator will display "Ans∗"; press [3] and [enter].

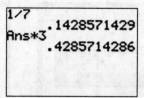

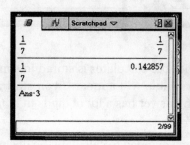 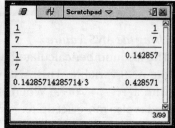

The ANS variable can also help you do calculations in stages. To calculate $x = \dfrac{-1 + \sqrt{5}}{2}$ without using so many parentheses as before, it can be done by first calculating $-1 + \sqrt{5}$ and then pressing $[\div]$ and $[2]$ and Ans will appear automatically.

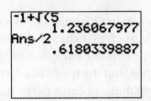

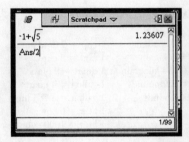
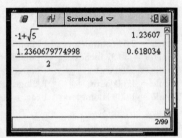

The ANS variable can also be accessed by pressing [2ND] and $[-]$ at the bottom right of the calculator. If after calculating the decimal equivalent of $\dfrac{1}{7}$ you wanted to subtract $\dfrac{1}{7}$ from 5, for the TI-84 press [5], $[-]$, [2ND], [ANS], and [ENTER]. For the TI-Nspire press [5], $[-]$, [ctrl], [ans], and [enter].

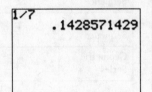
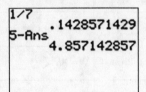

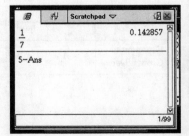

TIP 4
Use the Reference Sheet

Suggestion

- In the back of the Algebra I Regents booklet is a reference sheet that contains 17 conversion facts, such as inches to centimeters and quarts to pints, and also 17 formulas. Many of these conversion facts and formulas will not be needed for an individual test, but the quadratic formula and the arithmetic sequence formula are the two that will come in the handiest.

High School Math Reference Sheet

1 inch = 2.54 centimeters	1 kilometer = 0.62 mile	1 cup = 8 fluid ounces
1 meter = 39.37 inches	1 pound = 16 ounces	1 pint = 2 cups
1 mile = 5280 feet	1 pound = 0.454 kilogram	1 quart = 2 pints
1 mile = 1760 yards	1 kilogram = 2.2 pounds	1 gallon = 4 quarts
1 mile = 1.609 kilometers	1 ton = 2000 pounds	1 gallon = 3.785 liters
		1 liter = 0.264 gallon
		1 liter = 1000 cubic centimeters

Triangle	$A = \frac{1}{2}bh$	Pythagorean Theorem	$a^2 + b^2 = c^2$
Parallelogram	$A = bh$	Quadratic Formula	$x = \frac{-b \pm \sqrt{b^2 - 4ac}}{2a}$
Circle	$A = \pi r^2$	Arithmetic Sequence	$a_n = a_1 + (n - 1)d$
Circle	$C = \pi d$ or $C = 2\pi r$	Geometric Sequence	$a_n = a_1 r^{n-1}$
General Prisms	$V = Bh$	Geometric Series	$S_n = \frac{a_1 - a_1 r^n}{1 - r}$ where $r \neq 1$
Cylinder	$V = \pi r^2 h$	Radians	1 radian = $\frac{180}{\pi}$ degrees
Sphere	$V = \frac{4}{3}\pi r^3$	Degrees	1 degree = $\frac{\pi}{180}$ radians
Cone	$V = \frac{1}{3}\pi r^2 h$	Exponential Growth/Decay	$A = A_0 e^{k(t - t_0)} + B_0$
Pyramid	$V = \frac{1}{3}Bh$		

Key Algebra I Topics: A Quick Review and Practice

1. Properties of Algebra and Solving Linear Equations with Algebra

1.1 One-Step Algebra Problems

An algebra problem, like $x - 5 = 2$, is one that can be solved by changing both sides of the equation until the variable x is isolated. There are four main properties that can be used in solving algebra problems.

- The **addition property** of equality

  The equation $x - 5 = 2$ is solved by adding 5 to both sides of the equation. When you add to both sides of an equation, you are using the addition property of equality.

$$
\begin{array}{ll}
x - 5 = 2 & \text{The given equation} \\
+5 = +5 & \text{Addition property of equality} \\
x = 7 & x \text{ is isolated. The solution is 7.}
\end{array}
$$

- The **subtraction property** of equality

  The equation $x + 2 = 7$ is solved by subtracting 2 from both sides of the equation. When you subtract from both sides of an equation, you are using the subtraction property of equality.

$$
\begin{array}{ll}
x + 2 = 7 & \text{The given equation} \\
-2 = -2 & \text{Subtraction property of equality} \\
x = 5 & x \text{ is isolated. The solution is 5.}
\end{array}
$$

- The **division** property of equality

  The equation $2x = 10$ is solved by dividing both sides of the equation by 2. When you divide both sides of the equation, you are using the division property of equality.

  $$2x = 10 \qquad \text{The given equation}$$

  $$\frac{2x}{2} = \frac{10}{2} \qquad \text{Division property of equality}$$

  $$x = 5 \qquad x \text{ is isolated. The solution is 5.}$$

- The **multiplication** property of equality

  The equations $\frac{x}{5} = 3$ and $\frac{2}{3}x = 8$ can be solved by multiplying both sides of the equation by the same number.

  $$\frac{x}{5} = 3 \qquad \text{The given equation}$$

  $$5 \cdot \frac{x}{5} = 5 \cdot 3 \qquad \text{Multiplication property of equality}$$

  $$x = 15 \qquad x \text{ is isolated. The solution is 15.}$$

  $$\frac{2}{3}x = 8 \qquad \text{The given equation}$$

  $$\frac{3}{2} \cdot \frac{2}{3}x = \frac{3}{2} \cdot 8 \qquad \text{Multiplication property of equality}$$

  $$x = 12 \qquad x \text{ is isolated. The solution is 12.}$$

1.2 Two-Step Algebra Problems

When an equation has the form $mx + b = y$, it takes two steps to solve for x. The first step is to eliminate the b, which is called the **constant**. The second step is to eliminate the m, which is called the **coefficient**. The b is eliminated with either the addition or the subtraction property of equality. The m is eliminated with either the division or the multiplication property of equality.

- For the equations $3x - 7 = 11$ and $\frac{2}{3}x + 5 = 11$

$3x - 7 = 11$	The given equation
$+7 = +7$	Addition property of equality
$\frac{3x}{3} = \frac{18}{3}$	Division property of equality
$x = 6$	x is isolated. The solution is 6.

$\frac{2}{3}x + 5 = 11$	The given equation
$-5 = -5$	Subtraction property of equality
$\frac{2}{3}x = 6$	
$\frac{3}{2} \cdot \frac{2}{3}x = \frac{3}{2} \cdot 6$	Multiplication property of equality
$x = 9$	x is isolated. The solution is 9.

1.3 Combining Like Terms Before Solving

If the equation has multiple x-terms or multiple constants, the equation should first be simplified by combining like terms. After all like terms have been combined, the question will usually be a two-step algebra problem and can be solved with the methods from Section 1.2.

$2x + 5 + 3x - 2 = 23$	The given equation
$2x + 3x + 5 - 2 = 23$	Terms are rearranged so the x-terms are together and the constants are together. This step is optional.
$5x + 3 = 23$	Like terms have been combined.
$-3 = -3$	Subtraction property of equality
$\frac{5x}{5} = \frac{20}{5}$	Division property of equality
$x = 4$	x is isolated. The solution is 4.

1.4 Variables on Both Sides of the Equation

When there are x-terms on both sides of the equation, the addition property of equality or the subtraction property of equality can be used to change the equation into one where the x-terms are all on the same side of the equation.

$5x - 3 = 12 + 2x$	The given equation
$-2x = -2x$	Subtraction property of equality eliminates the x-term from the right-hand side of the equation
$3x - 3 = 12$	
$+3 = +3$	Addition property of equality
$\dfrac{3x}{3} = \dfrac{15}{3}$	Division property of equality
$x = 5$	x is isolated. The solution is 5.

1.5 Equations with More Than One Variable

An equation with more than one variable can be solved the same way as an equation with just one variable. The **solution** will not be a number in these problems but, instead, an **expression** with variables and numbers in it.

$ax + 2 = c$	The given equation
$-2 = -2$	Subtraction property of equality
$ax = c - 2$	The c and the 2 cannot be combined since they are unlike terms
$\dfrac{ax}{a} = \dfrac{c - 2}{a}$	Division property of equality
$x = \dfrac{c - 2}{a}$	x is isolated. The solution for x is not a number but an expression in terms of c and a. The answer is $\dfrac{c - 2}{a}$.

Practice Exercises: Topic 1

1. Antonio started the question $2x + 1 = 11$ by writing $2x = 10$. Which property justifies this step?

(1) Commutative property of addition

(2) Distributive property of multiplication over addition

(3) Addition property of equality

(4) Subtraction property of equality

2. Mila used the multiplication property to justify the first step in solving an equation. The original equation was $\frac{x}{2} + 4 = 10$. What could the equation have been transformed into after this step?

(1) $\frac{x}{2} = 6$

(2) $\frac{x}{2} + 6 + 12$

(3) $x + 8 = 20$

(4) $x + 2 = 5$

3. What is the solution set for the equation $x - 6 = 7$?

(1) $\{1\}$

(2) $\{6\}$

(3) $\{11\}$

(4) $\{13\}$

4. What value of x makes the equation $3x + 7 = 22$ true?

(1) 1

(2) 3

(3) 5

(4) 7

5. Find the solution set for the equation $5(x + 4) = 35$.

(1) $\{1\}$

(2) $\{2\}$

(3) $\{3\}$

(4) $\{4\}$

6. Solve for d in terms of c, e, and f.

$cd - e = f$

(1) $\dfrac{f - e}{c}$

(3) $\dfrac{f}{c} + e$

(2) $\dfrac{f + e}{c}$

(4) $\dfrac{f}{c} - e$

7. Solve for m in terms of a, b, and c.

$b - ma = c$

(1) $\dfrac{b - c}{-a}$

(3) $\dfrac{c - b}{-a}$

(2) $(c - b) - a$

(4) $(c + b) - a$

8. Solve for r in terms of c and π.

$c = 2\pi r$

(1) $\dfrac{2c}{\pi}$

(3) $\dfrac{c}{2\pi}$

(2) $\dfrac{2\pi}{c}$

(4) $\dfrac{2}{c\pi}$

Solutions for Practice Exercises: Topic 1

1. The first step of the process is to subtract 1 from each side of the equation. This is called the subtraction property of equality.

   The correct choice is **(4)**.

2. Though it is more common to begin this question by subtracting 4 from both sides of the equation, in this case she does it by multiplying both sides of the equation by 2. The left-hand side becomes $x + 8$ and the right-hand side becomes 20.

   The correct choice is **(3)**.

3. Isolate the x by adding 6 to both sides of the equation. The equation then becomes $x = 13$.

   The correct choice is **(4)**.

4. Subtract 7 from both sides of the equation to get $3x = 15$. Divide both sides of the equation by 3 to get $x = 5$.

   The correct choice is **(3)**.

5. One way to solve this equation is to first distribute the 5 through the left-hand side to get the equation $5x + 20 = 35$, then subtract 20 from both sides of the equation to get $5x = 15$, and finally divide both sides of the equation by 5 to get $x = 3$. Another way is to first divide both sides of the equation by 5 to get $x + 4 = 7$ and then subtract 4 from both sides of the equation to get $x = 3$.

   The correct choice is **(3)**.

6. First add e to both sides of the equation to get $cd = f + e$. Then divide both sides of the equation by c to get $d = \dfrac{f + e}{c}$.

   The correct choice is **(2)**.

7. First subtract b from both sides of the equation to get $-ma = c - b$. Then divide both sides by $-a$ to get $m = \dfrac{c - b}{-a}$.

   The correct choice is **(3)**.

8. Divide both sides of the equation by 2π to get $\dfrac{c}{2\pi} = r$.

   The correct choice is **(3)**.

2. Polynomial Arithmetic

2.1 Classifying Polynomials

A **polynomial** is an expression like $2x + 5$ or $3x^2 - 5x + 3$. The **terms** of a polynomial are separated by $+$ or $-$ signs. The polynomial $2x + 5$ has two terms. The polynomial $3x^2 - 5x + 3$ has three terms. The terms of a polynomial have a **coefficient** and a **variable part**. The term $3x^2$ has a coefficient of 3 and a variable part of x^2. A term with no variable part is called a **constant**.

- A polynomial with three terms is called a **trinomial**.
- A polynomial with two terms is called a **binomial**.
- A polynomial with one term is called a **monomial**.

2.2 Multiplying and Dividing Monomials

- To *multiply* one monomial by another, multiply the coefficients and multiply the variable parts by adding the exponents on the same variables.

$$8x^3 \cdot 2x$$

Multiply the coefficients $8 \cdot 2 = 16$.

Multiply the variable parts by adding the exponents $x^3 \cdot x^1 = x^4$. The solution is $16x^4$.

- To *divide* one monomial by another, divide the coefficients and divide the variable parts by subtracting the exponents on the same variables.

$$8x^3 \div 2x$$

Divide the coefficients. $8 \div 2 = 4$.

Divide the variable parts by subtracting the exponents

$$x^3 \div x^1 = x^2$$

The solution is $4x^2$.

This question can also be expressed as $\dfrac{8x^3}{2x} = 4x^2$.

2.3 Combining Like Terms

Like terms are terms that have the same variable part. For example, $3x^2$ and $2x^2$ are like terms because the variable part for both is x^2. Like terms can be added or subtracted by adding or subtracting the coefficients and by not changing the variable part.

$$3x^2 + 2x^2 = 5x^2$$
$$3x^2 - x^2 = 3x^2 - 1x^2 = 2x^2$$

- To simplify $2x + 3 + 4x - 5$, combine the like terms with the variable part of x. $2x + 4x = 6x$. Also combine the constants $3 - 5 = -2$. This expression simplifies to $6x - 2$.

If terms are not like terms, they cannot be combined by adding or subtracting. $3x^2 + 5x^3$ cannot be combined because the exponents are different, so they are not like terms.

2.4 Multiplying Monomials and Polynomials

- To *multiply* a polynomial by a monomial, use the **distributive property**.

$$2(3x + 5) = 2 \cdot 3x + 2 \cdot 5 = 6x + 10$$

This works for more complex monomials and polynomials also.

$$2x^2(5x^2 - 7x + 3) = 2x^2 \cdot 5x^2 + 2x^2 \cdot -7x + 2x^2 \cdot 3$$
$$= 10x^4 - 14x^3 + 6x^2$$

2.5 Adding and Subtracting Polynomials

- To *add* two polynomials, remove the parentheses from both and combine like terms.

$$(5x + 2) + (3x - 4) = 5x + 2 + 3x - 4 = 8x - 2$$

- To *subtract* two polynomials, remove the parentheses of the polynomial on the left, then negate all the terms of the polynomial on the right, and remove the parentheses before combining like terms.

$$(5x + 2) - (3x - 4) = 5x + 2 - 3x + 4 = 2x + 6$$

2.6 Multiplying Binomials

- To *multiply* binomials, use the **FOIL** process.

$$(2x + 3)(5x - 2)$$

The F stands for firsts. Multiply $2x \cdot 5x$, the first term in each of the parentheses: $10x^2$.

The O stands for outers. Multiply $2x \cdot -2$, the terms on the far left and on the far right: $-4x$.

The I stands for inners. Multiply $3 \cdot 5x$, the two terms in the middle: $15x$.

The L stands for lasts. Multiply $3 \cdot -2$, the second term in each of the parentheses: -6.

These four answers become $10x^2 - 4x + 15x - 6$. Combine like terms to get $10x^2 + 11x - 6$ in simplified form.

2.7 Factoring Polynomials

Factoring a number, like 15, is when two numbers are found that can be multiplied to become that number: $15 = 3 \cdot 5$. Factoring polynomials is more involved, and there are certain patterns to be aware of.

- **Greatest Common Factor Factoring**

The terms of some polynomials have a greatest common factor. This can be factored out like a reverse use of the distributive property.

In $6x^2 + 8x$, the terms have a common factor of $2x$. Write $2x$ outside the parentheses, and divide each term by $2x$ to determine what goes inside the parentheses.

$$2x(3x + 4)$$

- **Difference of Perfect Squares Factoring**

The expression $a^2 - b^2$ can be factored into $(a - b)(a + b)$. This works anytime both terms of a binomial are perfect squares and there is a minus sign between the two terms.

$$x^2 - 9 = x^2 - 3^2 = (x - 3)(x + 3)$$

- **Reverse FOIL**

A **trinomial** like $x^2 + 8x + 15$ can be factored if there are two numbers that have a sum equal to the coefficient of the x-term, 8, and a product equal to the constant, 15. Since $3 + 5 = 8$ and $3 \cdot 5 = 15$,

$$x^2 + 8x + 15 = (x + 3)(x + 5)$$

For $x^2 + 3x - 10$, the numbers that have a sum of 3 and a product of -10 are -2 and 5.

$$x^2 + 3x - 10 = (x - 2)(x + 5)$$

2.8 More Complicated Factoring

Sometimes none of the factoring patterns seem to match the polynomial that needs to be factored. When this happens, see if it is possible to rewrite it in a way that resembles the pattern better.

- The polynomial $x^4 - 9$ can be rewritten as $(x^2)^2 - 3^2$, which now has the difference of perfect squares pattern.

$$x^4 - 9 = (x^2)^2 - 3^2 = (x^2 - 3)(x^2 + 3)$$

- The polynomial $x^4 + 8x^2 + 15$ can be rewritten as

$$(x^2)^2 + 8(x^2) + 15$$
$$x^4 + 8x^2 + 15 = (x^2)^2 + 8(x^2) + 15 = (x^2 + 3)(x^2 + 5)$$

Practice Exercises: Topic 2

1. Classify this polynomial $5x^2 + 3$.

   (1) Monomial

   (2) Binomial

   (3) Trinomial

   (4) None of the above

2. Classify this polynomial $7x^2 - 3x + 2$.

   (1) Monomial

   (2) Binomial

   (3) Trinomial

   (4) None of the above

3. Multiply $3x^3 \cdot 4x^5$.

   (1) $7x^8$

   (2) $7x^{15}$

   (3) $12x^8$

   (4) $12x^{15}$

4. Which expression is equivalent to $2x^2 + 5x^2$?

   (1) $10x^2$

   (2) $7x^2$

   (3) $7x^4$

   (4) The expression cannot be simplified any further.

5. Simplify $6x(2x + 3)$.

   (1) $8x^2 + 18x$

   (2) $12x^2 + 18x$

   (3) $12x^2 + 3$

   (4) $20x$

6. Simplify $3x(5x^2 - 2x + 3)$.

   (1) $15x^3 - 2x + 3$

   (2) $15x^3 - 6x^2 + 9x$

   (3) $5x^2 + x + 3$

   (4) $15x^3 + 6x^2 - 9x$

7. Simplify $(3x - 4) - (5x - 3)$. $-15x^2 + 9x + 20x + 12$
   - (1) $-2x - 7$
   - (2) $-2x - 1$
   - (3) $2x - 1$
   - (4) $2x - 7$

8. $(2x + 3)(3x - 1) = 6x^2 - 2x + 9x$
   - (1) $6x^2 - 3$
   - (2) $6x^2 + 11x - 3$
   - (3) $6x^2 + 7x - 3$
   - (4) $6x^2 - 7x - 3$

9. Factor $x^2 - 2x - 15$.
   - (1) $(x - 3)(x - 5)$
   - (2) $(x + 3)(x - 5)$
   - (3) $(x - 15)(x + 1)$
   - (4) $(x + 15)(x - 1)$

10. Factor $x^4 + 7x^2 + 12$.
    - (1) $(x^2 + 6)(x^2 + 2)$
    - (2) $(x^2 - 3)(x^2 - 4)$
    - (3) $(x^2 + 3)(x^2 + 4)$
    - (4) This cannot be factored.

Solutions for Practice Exercises: Topic 2

1. There are two terms, $5x^2$ and 3, separated by a + sign. A polynomial with two terms is called a binomial.

   The correct choice is **(2)**.

2. There are three terms, $7x^2$, $3x$, and 2, separated by − and + signs. A polynomial with three terms is called a trinomial.

   The correct choice is **(3)**.

3. To multiply two monomials, first multiply the coefficients: $3 \cdot 4 = 12$. Then multiply the variable parts. Remember that when multiplying variables, you add the exponents: $x^3 \cdot x^5 = x^{(3+5)} = x^8$. The solution is $12x^8$.

   The correct choice is **(3)**.

4. Since these are like terms with the variable part x^2, they can be combined. The solution will also have a variable part of x^2 with a coefficient equal to the sum of the two coefficients. Since $2 + 5 = 7$, $2x^2 + 5x^2 = 7x^2$.

   The correct choice is **(2)**.

5. Using the distributive property gives $6x \cdot 2x + 6x \cdot 3 = 12x^2 + 18x$.

   The correct choice is **(2)**.

6. Using the distributive property results in $3x \cdot 5x^2 + 3x(-2x) + 3x(3) = 15x^3 - 6x^2 + 9x$.

   The correct choice is **(2)**.

7. Distribute the − sign through the parentheses on the right. The expression becomes $3x - 4 - 5x + 3$. Combine like terms to get $-2x - 1$.

   The correct choice is **(2)**.

8. Use the FOIL process. The firsts are $2x \cdot 3x = 6x^2$. The outers are $2x \cdot (-1) = -2x$. The inners are $3 \cdot 3x = 9x$. The lasts are $3 \cdot (-1) = -3$. Combine these four terms to get $6x^2 - 2x + 9x - 3$. Combine like terms to get $6x^2 + 7x - 3$.

The correct choice is **(3)**.

9. To factor this quadratic trinomial, find two numbers that have a product of -15 and a sum of -2. The numbers are $+3$ and -5. This factors into $(x+3)(x-5)$.

The correct choice is **(2)**.

10. This trinomial can be written as $(x^2)^2 + 7(x^2) + 12$. It has the same structure, then, as a quadratic trinomial and can be factored by finding two numbers that have a product of 12 and a sum of 7. The two numbers are $+3$ and $+4$. This factors into $(x^2 + 3)(x^2 + 4)$.

The correct choice is **(3)**.

3. Quadratic Equations

3.1 Methods of Solving Quadratic Equations

A **quadratic equation** is an equation that can be written in the form $ax^2 + bx + c = 0$. For example, $x^2 + 4x - 5 = 0$ is a quadratic equation. There are three ways to solve a quadratic equation.

1. Solve by factoring. If possible, factor the left-hand side of the equation.

$$x^2 + 4x - 5 = 0$$
$$(x + 5)(x - 1) = 0$$

Since the only way that two things can have a product of zero is if at least one of them is zero, this means that either $(x + 5)$ or $(x - 1)$ must equal zero.

$$
\begin{array}{lcl}
x + 5 = 0 & \text{or} & x - 1 = 0 \\
-5 = -5 & & +1 = +1 \\
x = -5 & \text{or} & x = 1
\end{array}
$$

$x = -5$ or $x = 1$ are the solutions to this quadratic equation.

2. Solve by completing the square. First eliminate the constant from the left-hand side by adding or subtracting.

$$x^2 + 4x - 5 = 0$$
$$+5 = +5$$
$$x^2 + 4x = 5$$

Next, *divide* the coefficient of the x by 2, square that answer, and add that number to both sides of the equation.

$$\left(\frac{4}{2}\right)^2 = 2^2 = 4$$

$$x^2 + 4x + 4 = 5 + 4$$
$$x^2 + 4x + 4 = 9$$

The left-hand side of the equation will factor.

$$(x + 2)(x + 2) = 9$$
$$(x + 2)^2 = 9$$

Take the square root of both sides of the equation, putting a $\pm$ in front of the square root on the right-hand side.

$$\sqrt{(x+2)^2} = \pm\sqrt{9}$$
$$x + 2 = \pm 3$$
$$-2 = -2$$
$$x = -2 \pm 3$$
$$x = -2 + 3 \text{ or } x = -2 - 3$$
$$x = 1 \text{ or } x = -5$$

3. Solve with the quadratic formula. Any quadratic equation of the form $ax^2 + bx + c = 0$ can be solved with the equation

$$x = \frac{-b \pm \sqrt{b^2 - 4ac}}{2a}$$

For $x^2 + 4x - 5 = 0$, $a = 1$, $b = 4$, $c = -5$.

$$x = \frac{-4 \pm \sqrt{4^2 - 4 \cdot 1 \cdot (-5)}}{2 \cdot 1} = \frac{-4 \pm \sqrt{16 + 20}}{2} = \frac{-4 \pm \sqrt{36}}{2} = \frac{-4 \pm 6}{2}$$

$$x = \frac{-4 + 6}{2} = \frac{2}{2} = 1 \text{ or } x = \frac{-4 - 6}{2} = -\frac{10}{2} = -5$$

3.2 The Relationship between Factors and Zeros

If a quadratic equation has factors $(x - p)$ and $(x - q)$, then the roots of the equation are p and q. If the equation has roots (or zeros) p and q, the factors are $(x - p)$ and $(x - q)$.

For example, the factors of the equation $x^2 - 7x + 10$ are $(x - 5)$ and $(x - 2)$. Therefore, the zeros of the equation are 5 and 2.

If the roots of a quadratic equation are -2 and 8, then the factors are $(x - (-2))$ and $(x - 8)$. The $(x - (-2))$ can be expressed as $(x + 2)$.

3.3 Using the Discriminant to Determine the Number of Unique Answers to a Quadratic Equation

Although most quadratic equations have two different answers, some have just one answer and some have no real answers. If you are asked how many real answers a quadratic equation, $ax^2 + bx + c = 0$, has, there is a quick way to determine this: Calculate the **discriminant**, $D = b^2 - 4ac$.

When you calculate the discriminant, one of three things can happen:

- If $b^2 - 4ac > 0$, there are two real answers. For example, $x^2 + 6x - 2 = 0$ has two real solutions because $D = b^2 - 4ac = 6^2 - 4 \cdot 1 \cdot (-2) = 36 + 8 = 44 > 0$.

- If $b^2 - 4ac = 0$, there is one real answer. For example, $x^2 + 6x + 9 = 0$ has one real solution because $D = b^2 - 4ac = 6^2 - 4 \cdot 1 \cdot 9 = 36 - 36 = 0$.

- If $b^2 - 4ac < 0$, there are no real answers. For example, $x^2 + 6x + 10 = 0$ has no real solutions because $D = b^2 - 4ac = 6^2 - 4 \cdot 1 \cdot 10 = 36 - 40 = -4 < 0$.

3.4 Word Problems Involving Quadratic Equations

Some real-world scenarios can be modeled with quadratic equations.

- **Area Problems**

    The width of a rectangle is 3 units more than the length. If the area of the rectangle is 70 square units, what are the length and width of the rectangle?

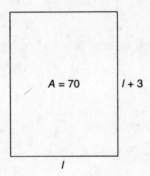

    Since area is length times width, for this scenario:

$$70 = l \cdot w$$
$$70 = l \cdot (l + 3)$$
$$70 = l^2 + 3l$$
$$0 = l^2 + 3l - 70$$

    Any of the methods can be used to solve for the answers $l = -10$ and $l = 7$. Since the length must be positive, the answer is that the length is 7 units and the width is $7 + 3 = 10$ units.

■ **Projectile Problems**

The height of a projectile after t seconds can be modeled with a quadratic equation. If the equation for the height of a baseball is $h = -16t^2 + 48t + 64$, when will the baseball land on the ground?

When the ball lands on the ground, the height will be 0.
$0 = -16t^2 + 48t + 64$ is the equation.

This can be solved by any of the methods to find the answers $t = -1$ and $t = 4$. Since the amount of time must be positive, the solution is 4 seconds.

Practice Exercises: Topic 3

1. Find all solutions to $(x + 2)^2 = 64$. $x+2=\pm 8 \quad x=2\pm 8$

   (1) $\sqrt{62}, -\sqrt{62}$ ✗

   (2) $6, -10$

   (3) 6

   (4) -10 ✗

2. Use completing the square to find both solutions for x in the equation $x^2 + 8x + 16 = 9$. $(x+4)^2 = \pm\sqrt 9$

   (1) $-1, -7$

   (2) $-1, -2$

   $x+4 = \pm 3$

   $x = -4 \pm 3$

   (3) $-2, -3$

   (4) $-3, -4$

3. Find all solutions to $x^2 + 6x = 0$.

   (1) 0

   (2) -6

   $x(x+6) = 0$

   (3) $0, -6$

   (4) $0, 6$

4. Solve $x^2 + 10x + 24 = 0$ by factoring.

   (1) $4, -6$

   (2) $-4, -6$

   (3) $-4, 6$

   (4) $4, 6$

5. If the roots of an equation are 3 and -6, what could the equation be?

   (1) $(x - 3)(x - 6) = 0$

   (2) $(x - 3)(x + 6) = 0$

   (3) $(x + 3)(x - 6) = 0$

   (4) $(x + 3)(x + 6) = 0$

6. If the roots of a polynomial are 1 and -8, what could be the factors?

   (1) $(x - 1)$ and $(x + 8)$

   (2) $(x - 1)$ and $(x - 8)$

   (3) $(x + 1)$ and $(x + 8)$

   (4) $(x + 1)$ and $(x - 8)$

7. Solve $x^2 + 4x - 7 = 0$ with the quadratic formula.

   (1)   1.3

   (2) $2 \pm \sqrt{11}$

   (3) $2 \pm \sqrt{12}$

   (4) $-2 \pm \sqrt{11}$

8. How many unique real answers does the equation $2x^2 - 8x + 3 = 0$ have?

   (1)   1

   (2)   2

   (3)   3

   (4)   0

9. The height of a projectile in feet at time t is determined by the equation $h = -16t^2 + 128t + 320$. At what time will the projectile be 560 feet high?

   (1)   4 seconds

   (2)   5 seconds

   (3)   6 seconds

   (4)   7 seconds

10. The height of a projectile in feet at time t is determined by the equation $h = -16t^2 + 112t + 128$. At what time will the projectile be 0 feet high?

   (1)   5 seconds

   (2)   6 seconds

   (3)   7 seconds

   (4)   8 seconds

Solutions for Practice Exercises: Topic 3

1. Take the square root of both sides to get $x + 2 = \pm 8$. The equation $x + 2 = 8$ has the solution $x = 6$. The equation $x + 2 = -8$ has the solution $x = -10$.

   The correct choice is **(2)**.

2. Since the constant, 16, is already equal to the square of half the coefficient $\left(\frac{8}{2}\right)^2$, the left-hand side of the equation is already a perfect square trinomial. It can be factored as $(x + 4)^2 = 9$. Then take the square root of both sides to get $x + 4 = \pm 3$. The equation $x + 4 = 3$ has the solution $x = -1$. The equation $x + 4 = -3$ has the solution $x = -7$.

   The correct choice is **(1)**.

3. Factor out an x to get the equation $x(x + 6) = 0$. This equation is true when either $x = 0$ or when $x + 6 = 0$, which becomes $x = -6$.

   The correct choice is **(3)**.

4. The two numbers that have a product of 24 and a sum of 10 are $+4$ and $+6$. The factors are $(x + 4)(x + 6)$. The solutions to the equation $(x + 4)(x + 6) = 0$ are when $x + 4 = 0$, which becomes $x = -4$, and also when $x + 6 = 0$, which becomes $x = -6$.

   The correct choice is **(2)**.

5. When a is a root of an equation, $(x - a)$ is a factor. If the roots are 3 and -6, the factors can be $(x - 3)$ and $(x - (-6)) = (x + 6)$.

   The correct choice is **(2)**.

6. If 1 is a root, $(x - 1)$ is a factor. If -8 is a root, $(x - (-8)) = (x + 8)$ is a factor.

   The correct choice is **(1)**.

7. $a = 1, b = 4, c = -7.$

$$x = \frac{-4 \pm \sqrt{4^2 - 4(1)(-7)}}{2} = \frac{-4 \pm \sqrt{16 + 28}}{2} = \frac{-4 \pm \sqrt{44}}{2}$$

$$= \frac{-4 \pm 2\sqrt{11}}{2} = -2 \pm \sqrt{11}$$

The correct choice is **(4)**.

8. Calculate $D = b^2 - 4ac = (-8)^2 - 4 \cdot 2 \cdot 3 = 64 - 24 = 40$. Since $D = 40 > 0$, there are two unique real answers.

The correct choice is **(2)**.

9. When 560 is substituted for h, the equation becomes $560 = -16t^2 + 128t + 320$. Subtract 560 from both sides of the equation to get $0 = -16t^2 + 128t - 240$. Divide both sides by -16 to get $0 = t^2 - 8t + 15$. The right-hand side factors, and the equation becomes $0 = (t - 3)(t - 5)$ with solutions $t = 3$ and $t = 5$. Of the choices listed, only 5 seconds is correct.

The correct choice is **(2)**.

10. Substituting $h = 0$ into the equation gives $0 = -16t^2 + 112t + 128$. Divide both sides by -16 to get $0 = t^2 - 7t - 8$. The right side factors, so the equation becomes $0 = (t - 8)(t + 1)$. The solutions are $t = 8$ and $t = -1$. Since the amount of time must be positive, the -1 is rejected.

The correct choice is **(4)**.

4. Systems of Linear Equations

4.1 What Is a System of Linear Equations?

A **system of equations** is a set of two equations that each have two variables. The system is *linear* if there are no exponents greater than 1 on any of the variables.

$$2x + 3y = 21$$
$$5x - 2y = 5$$

is a system of linear equations.

The solution to a system of equations is the set of **ordered pairs** that satisfy both equations at the same time. For the system above, the solution is the ordered pair $(3, 5)$ since

$$2(3) + 3(5) = 6 + 15 = 21$$
$$5(3) - 2(5) = 15 - 10 = 5$$

There are two main techniques for solving systems of equations.

4.2 Solving a System of Equations with the Substitution Method

If one of the two equations is in the form $y = mx + b$, use the **substitution method**.

$$y = 2x + 3$$
$$2x + 4y = 42$$

- Substitute the expression $2x + 3$ for the y in the second equation.

$$2x + 4(2x + 3) = 42$$

Solve for x.

$$2x + 8x + 12 = 42$$
$$10x + 12 = 42$$
$$-12 = -12$$
$$\frac{10x}{10} = \frac{30}{10}$$
$$x = 3$$

- Substitute 3 for x into either of the original equations, and solve for y.

$$y = 2(3) + 3$$
$$y = 6 + 3$$
$$y = 9$$

The solution is the ordered pair $(3, 9)$.

4.3 Solving a System of Equations with the Elimination Method

When both equations are in the form $ax + by = c$, use the **elimination method**. The elimination method is when the two equations are combined in a way that eliminates one of the variables.

In the system of equations below, the y-term will be eliminated if the two equations are added. When the coefficient of one of the variables in one equation is the same number with the opposite sign of the same variable in the other equation, add the equations to eliminate that variable. In this case, the $+2y$ has the opposite coefficient as the $-2y$.

$$3x + 2y = 19$$
$$+4x - 2y = 16$$
$$\frac{7x}{7} = \frac{35}{7}$$
$$x = 5$$

■ Substitute 5 for x into either of the original equations

$$3(5) + 2y = 19$$
$$15 + 2y = 19$$
$$-15 = -15$$
$$\frac{2y}{2} = \frac{4}{2}$$
$$y = 2$$

The solution is the ordered pair $(5, 2)$.

For some systems of equations, one or both of the equations need to be changed so the elimination will happen.

In the system

$$2x - 4y = 4$$
$$3x + 2y = 14$$

since 4 is a multiple of 2, multiply both sides of the bottom equation by 2 and the coefficients of the y-variables will be the same number with opposite signs.

$$2x - 4y = 4$$
$$2(3x + 2y) = 2(14)$$

$$2x - 4y = 4$$
$$\underline{+6x + 4y = 28}$$
$$\frac{8x}{8} = \frac{32}{8}$$
$$x = 4$$

$$2(4) - 4y = 4$$
$$8 - 4y = 4$$
$$-8 = -8$$
$$\frac{-4y}{-4} = \frac{-4}{-4}$$
$$y = 1$$

The solution is the ordered pair (4, 1).

In the system

$$3x - 5y = 17$$
$$2x + 4y = 4$$

■ To eliminate the y, determine the least common multiple of 4 and 5, which is 20. Multiply both sides of both equations so that one of the coefficients on the y is -20 and the other is $+20$. To do this, multiply both sides of the top equation by 4 and multiply both sides of the bottom equation by 5.

$$4(3x - 5y) = 4(17)$$
$$5(2x + 4y) = 5(4)$$

$$12x - 20y = 68$$
$$\underline{+10x + 20y = 20}$$
$$\frac{22x}{22} = \frac{88}{22}$$
$$x = 4$$

- Substitute 4 for x in either of the original equations, and solve for y.

$$3(4) - 5y = 17$$
$$12 - 5y = 17$$
$$-12 = -12$$
$$\frac{-5y}{-5} = \frac{5}{-5}$$
$$y = -1$$

The solution to the system of equations is the ordered pair $(4, -1)$.

4.4 Word Problems Involving Systems of Equations

Some real-world scenarios can be modeled with a system of linear equations. Here is a typical example.

If five slices of pizza and three drinks cost $21 and if two slices of pizza and five drinks cost $16, how much is it for just one slice of pizza?

- Let x be the cost of a slice of pizza and y be the cost of a drink.

The system of equations is

$$5x + 3y = 21$$
$$2x + 5y = 16$$

- To eliminate the y, make the $3y$ and the $5y$ into $15y$ and $-15y$, respectively, by multiplying both sides of the top equation by 5 and both sides of the bottom equation by -3. Then add the two equations, and solve for x.

$$5(5x + 3y) = 5(21)$$
$$- 3(2x + 5y) = -3(16)$$

$$25x + 15y = 105$$
$$\underline{+-6x - 15y = -48}$$
$$\frac{19x}{19} = \frac{57}{19}$$
$$x = 3$$

Since the question just asked for the price of a slice of pizza, it is not necessary to also find the value of y. A slice of pizza costs $3.

Practice Exercises: Topic 4

1. $(2, 5)$ is a solution to which equation?

   (1) $x + 2y = 9$

   (2) $2x + y = 9$

   (3) $8x - y = 9$

   (4) $3x + 4y = 25$

2. Which equation has the same solution set as the equation $2x + 3y = 5$?

   (1) $8x + 12y = 20$

   (2) $8x + 12y = 15$

   (3) $6x + 9y = 12$

   (4) $4x + 6y = 8$

3. Solve the system of equations:

   $$3x + 2y = 17$$
   $$4x - 2y = 4$$

   (1) $(4, 3)$

   (2) $(5, 1)$

   (3) $(3, 4)$

   (4) $(1, 6)$

4. Solve the system of equations:

   $$y = 5x + 3$$
   $$2x + 6y = 50$$

   (1) $(1, 8)$

   (2) $(8, 1)$

   (3) $(-1, -8)$

   (4) $(-8, -1)$

5. Solve the system of equations:

   $$2x + 3y = -1$$
   $$-2x + 5y = -23$$

   (1) $(-4, 3)$

   (2) $(4, 3)$

   (3) $(4, -3)$

   (4) $(-4, -3)$

6. Solve the system of equations:

$$y = 3x - 2$$
$$4x - 2y = -4$$

   (1) $(10, 4)$ (3) $(-10, -4)$

   (2) $(4, 10)$ (4) $(-4, -10)$

7. In order to eliminate the x from this system of equations,

$$12x - 3y = 21$$
$$-2x + 6y = 2$$

you could

   (1)   multiply both sides of the first equation by 2.

   (2)   multiply both sides of the second equation by 6.

   (3)   multiply both sides of the first equation by -2.

   (4)   multiply both sides of the second equation by $1/2$.

8. Solve the system of equations:

$$8x - 2y = 28$$
$$4x + 3y = 6$$

   (1) $(3, 2)$ (3) $(-3, -2)$

   (2) $(-3, 2)$ (4) $(3, -2)$

9. Which system of equations can be used to model the following scenario?

There are 50 animals. Some of the animals have two legs, and the rest of them have four legs. In total there are 172 legs.

   (1) $x + y = 172$ (3) $y + 50 = x$

         $2x + 4y = 50$ $4y + 172 = 2x$

   (2) $x + 50 = y$ (4) $x + y = 50$

         $2x + 172 = 4y$ $2x + 4y = 172$

10. A pet store has 30 animals. Some are cats, and the rest are dogs. The cats cost $50 each. The dogs cost $100 each. If the total cost for all 30 animals is $1,900, how many cats are there?

   (1)   8   (3)   22

   (2)   20   (4)   24

Solutions for Practice Exercises: Topic 4

1. Substitute 2 for x and 5 for y into each of the choices. Choice (1) becomes $2 + 2(5) = 9$, which is not true. Choice (2) becomes $2(2) + 5 = 9$, which is true.

   The correct choice is **(2)**.

2. If both sides of an equation are multiplied by the same number, the new equation has the same solution set as the original equation. If both sides of the equation $2x + 3y = 5$ are multiplied by 4, it becomes $8x + 12y = 20$, which is choice (1). The other choices could be obtained by multiplying the right-hand side of the original equation by one number and the left-hand side of the original equation by another number, which does not produce an equation with the same solution set as the original.

   The correct choice is **(1)**.

3. Since the top equation has a $+2y$ and the bottom equation has a $-2y$, the equations can be added together. The y-terms will drop out, leading to the equation $7x = 21$ or $x = 3$. To solve for y, substitute 3 for x into either of the original equations, like $3(3) + 2y = 17$, $9 + 2y = 17$, $2y = 8$, which leads to $y = 4$.

   The correct choice is **(3)**.

4. Since the y is isolated in the top equation, substitute $5x + 3$ for y in the bottom equation. It becomes $2x + 6(5x + 3) = 50$, $2x + 30x + 18 = 50$, $32x + 18 = 50$, $32x = 32$, $x = 1$. Only one choice has an x-coordinate of 1.

   The correct choice is **(1)**.

5. Since the top equation has a $+2x$ and the bottom equation has a $-2x$, the equations can be added together to get $8y = -24$, $y = -3$. Substitute -3 for y into either original equation to solve for x: $2x + 3(-3) = -1$, $2x - 9 = -1$, $2x = 8$, $x = 4$.

   The correct choice is **(3)**.

6. Since the y is isolated in the top equation, substitute $3x - 2$ for y in the bottom equation: $4x - 2(3x - 2) = -4$, $4x - 6x + 4 = -4$, $-2x + 4 = -4$, $-2x = -8$, $x = 4$. To solve for y, substitute 4 for x into either original equation: $y = 3(4) - 2 = 12 - 2 = 10$.

   The correct choice is **(2)**.

7. The x will be eliminated after combining two equations when the coefficient of the x in one of the equations is the opposite of the coefficient of the x in the other equation. For choice (1), if both sides of the top equation are multiplied by 2, it would become $24x - 6y = 42$. $+24$ is not the opposite of -2. For choice (2), if both sides of the second equation are multiplied by 6, it becomes $-12x + 36y = 12$. Since -12 is the opposite of $+12$, this is the best answer.

   The correct choice is **(2)**.

8. Multiply both sides of the bottom equation by -2 to get $-8x - 6y = -12$. Add this to the top equation to eliminate the x and get $-8y = 16$. Divide both sides by -8 to get $y = -2$. Substitute -2 for y into one of the original equations: $8x - 2(-2) = 28$, $8x + 4 = 28$, $8x = 24$, $x = 3$.

   The correct choice is **(4)**.

9. If x is the number of two-legged animals and y is the number of four-legged animals, the number of animals is $x + y$ and the number of legs is $2x + 4y$. The system, then, is $x + y = 50$ and $2x + 4y = 172$.

   The correct choice is **(4)**.

10. If x is the number of dogs and y is the number of cats, the system of equations is

$$x + y = 30$$
$$100x + 50y = 1,900$$

   To eliminate the x, multiply both sides of the top equation by -100 to get $-100x - 100y = -3000$. Add this to the bottom equation to get $-50y = -1100$ or $y = 22$.

   The correct choice is **(3)**.

5. Graphs of Linear Equations

5.1 Graphing the Solution Set of a Linear Equation by Making a Table of Values

The equation $x + y = 10$ has an infinite number of ordered pairs that satisfy it. One way to organize the information before creating a graph is to make a table of values.

This is a chart with three ordered pairs satisfying the equation $x + y = 10$. For a linear equation, only two ordered pairs are needed, but it is wise to do an extra ordered pair in case one of your first two is incorrect.

x	y
2	8
3	7
9	1

Plot the ordered pair $(2, 8)$ on the coordinate plane by locating the point that is 2 units to the right of the y-axis and 8 units above the x-axis. One way to do this is to start at the origin point where the two axes intersect and move to the right 2 units from there and then up 8 units.

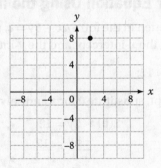

Do the same for the other two ordered pairs on the chart, $(3, 7)$ and $(9, 1)$.

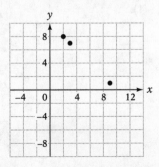

Draw a line through the three points. If the three points do not all lie on the same line, one of your ordered pairs is incorrect.

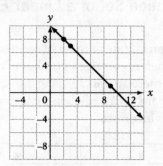

MATH FACTS

The line on a graph contains an infinite number of points. Each point corresponds to an ordered pair that is part of the solution set for the equation, and each ordered pair that is part of the solution set for the equation corresponds to a point on the line.

5.2 Graphing a Linear Equation Using the Intercept Method

When a linear equation is written in the form $ax + by = c$, the graph of the equation can be created quickly by finding the x-intercept and the y-intercept.

For the equation $2x - 3y = 12$:

- To find the y-intercept, substitute 0 for x and solve for y.

$$2(0) - 3y = 12$$
$$\frac{-3y}{-3} = \frac{12}{-3}$$
$$y = -4$$

The y-intercept is $(0, -4)$.

- To find the x-intercept, substitute 0 for y and solve for x.

$$2x - 3(0) = 12$$
$$2x = 12$$
$$x = 6$$

The x-intercept is $(6, 0)$.

- Plot $(0, -4)$ and $(6, 0)$ on a set of coordinate axes, and draw the line that passes through both points.

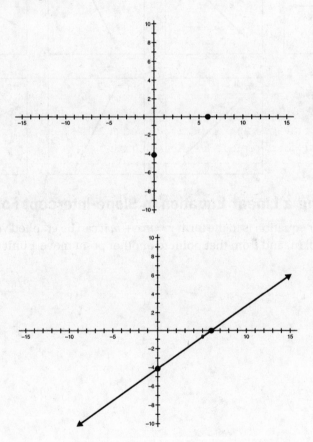

5.3 Calculating and Interpreting Slope

The *slope* of a line is a number that measures how steep it is. A horizontal line has a slope of 0. A line with a positive slope goes up as it goes to the right. A line with a negative slope goes down as it goes to the right. The variable used for slope is the letter *m*.

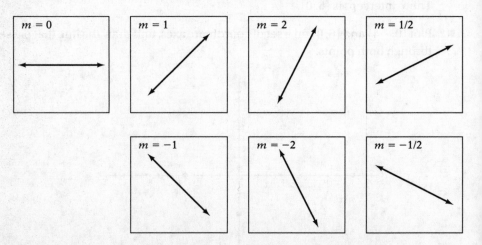

5.4 Graphing a Linear Equation in Slope-Intercept Form

When the linear equation is in the form $y = mx + b$, it can be graphed very quickly. The y-intercept is $(0, b)$, and from that point to another point move 1 unit to the right and m units up.

For the equation $y = 2x - 3$, the y-intercept is $(0, -3)$. From that point, move 1 unit to the right and 2 units up. Draw the line through the two points $(0, -3)$ and $(1, -1)$.

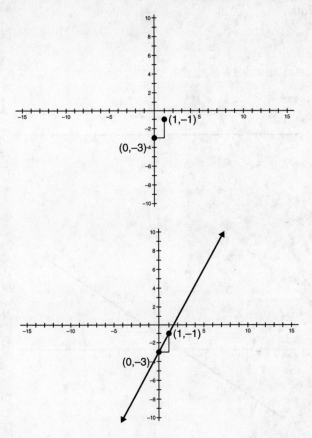

- If m is a fraction, n/d, it is more accurate to move d units to the right and n units up from the y-intercept.

■ For the equation $y = \frac{2}{3}x + 2$, the y-intercept is $(0, 2)$. From that point, move to the right and 2 up to get to the point $(3, 4)$. Draw a line through thes[e] two points.

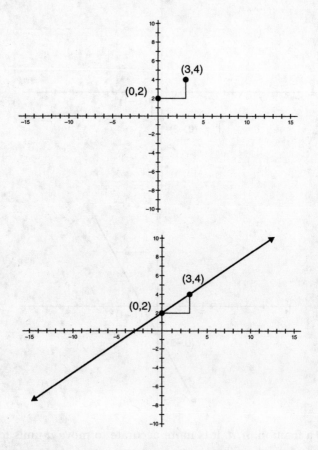

If the m-value is negative, move down instead of up to get from the y-intercept to th[e] next point.

5.5 Graphing Linear Equations on the Graphing Calculator

An equation in slope-intercept form can be quickly graphed on the graphing calculator.

- For the TI-84, press [Y=], enter the equation, and press [ZOOM] and [6].

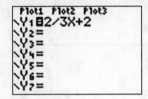 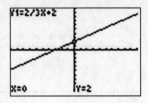

- For the TI-Nspire, go to the home screen and select [B] for the Graph Scratchpad. Enter the equation on the entry line after $f1(x)=$ and press [enter].

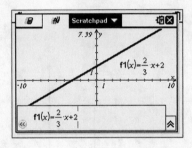

- If a linear equation with two variables is not originally in the slope-intercept form, $y = mx + b$, algebra can be used to rewrite it in slope-intercept form before graphing.

5.6 Equations of Vertical and Horizontal Lines

The graph of the equation $y = k$ is a horizontal line through the point $(0, k)$. The graph of the equation $x = h$ is a vertical line through the point $(h, 0)$.

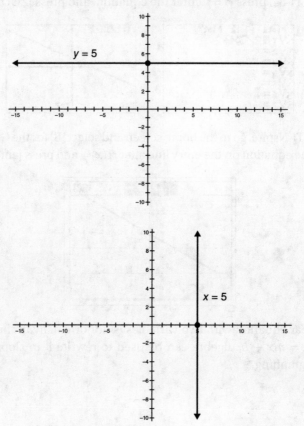

.7 Solving Systems of Linear Equations Graphically

he solution to a system of equations is the set of coordinates of the point where the raphs of the two lines intersect.

- The solution to the system of equations

$$y = 2x - 1$$

$$y = -\frac{2}{3}x + 7$$

can be found by carefully producing the two graphs on the same set of axes. The coordinates of the intersection point (3, 5) mean the solution to the system of equations is $x = 3$, $y = 5$.

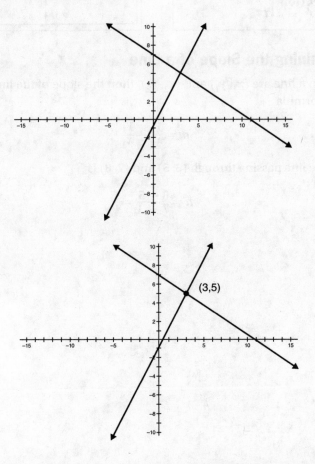

The graphing calculator can also determine the intersection point of two line On the TI-84, graph both lines and press [2ND], [TRACE], and [5] to find the inte section. On the TI-Nspire, graph both lines and press [menu], [6], and [4] to th intersection.

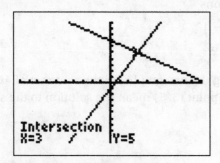
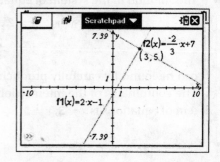

5.8 Determining the Slope of a Line

If two points on a line are (x_1, y_1) and (x_2, y_2), then the slope of the line can be deter mined by the formula

$$m = \frac{y_2 - y_1}{x_2 - x_1}$$

The slope of the line passing through $(3, 5)$ and $(7, 8)$ is

$$m = \frac{8 - 5}{7 - 3} = \frac{3}{4}$$

5.9 Interpreting the Slope of a Line

When a graph represents a real-world scenario, the slope is the rate that the x-value is changing with relation to the y-value. In a distance-time graph, the slope of a line segment corresponds to the speed that the moving object is traveling.

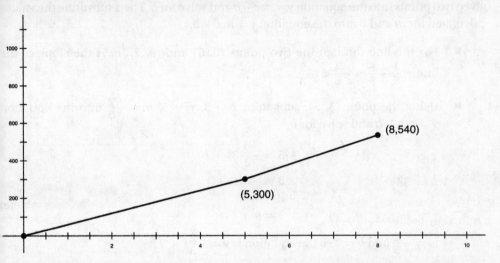

In the distance-time graph above, a car's time traveled is represented by the x-coordinate, and the distance it has covered at that time is represented by the y-coordinate. The slope of the segment connecting $(0, 0)$ and $(5, 300)$ is 60, so the car was traveling at a speed of 60 mph for the first 5 hours. The slope of the segment connecting $(5, 300)$ and $(8, 540)$ is 80, so the car was traveling at a speed of 80 mph for the last 3 hours.

5.10 Finding the Equation of a Line Through Two Given Points

If two points are known, use the slope-intercept formula to find the slope of the line. Substitute the value you calculated for m and the x- and y-coordinates of either of the given two points into the equation $y = mx + b$ and solve for b. Then substitute the values calculated for m and b into the equation $y = mx + b$.

- For the line through the two points (3, 5) and (9, 1), m is the slope of the line $= \dfrac{1-5}{9-3} = -\dfrac{4}{6} = -\dfrac{2}{3}$.

- Using the point (3, 5), substitute $x = 3$, $y = 5$, $m = -\dfrac{2}{3}$ into the equation $y = mx + b$ and solve for b.

$$5 = -\frac{2}{3}(3) + b$$
$$5 = -2 + b$$
$$+2 = +2$$
$$7 = b$$

$m = -\dfrac{2}{3}$ and $b = 7$ so the equation is $y = -\dfrac{2}{3}x + 7$.

Practice Exercises: Topic 5

1. What is the *x*-intercept of the graph of the solution set of the equation $2x + 5y = 20$?

(1) $(10, 0)$ (3) $(0, 10)$

(2) $(4, 0)$ (4) $(0, 4)$

2. This is a graph of the solution set of which equation?

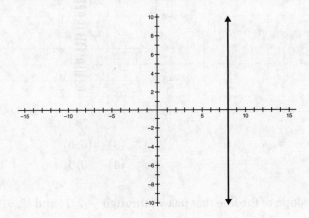

(1) $x = 8$ (3) $y = 8$

(2) $x = -8$ (4) $y = -8$

3. Below is the graph of the solution set of an equation. Based on this graph, which ordered pair does not seem to be part of the solution set of the equation $y = \frac{1}{3}x + 2$?

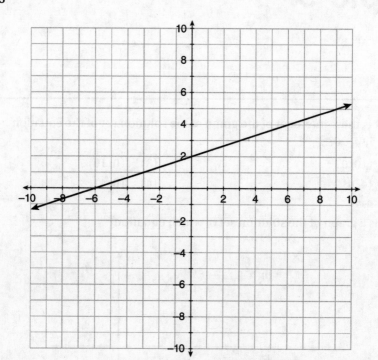

   (1) $(3, 3)$ (3) $(6, 8)$

   (2) $(6, 4)$ (4) $(9, 5)$

4. What is the slope of the line that passes through $(-2, 1)$ and $(8, 5)$?

   (1) $\frac{2}{5}$ (3) $\frac{5}{2}$

   (2) $-\frac{2}{5}$ (4) $-\frac{5}{2}$

5. Below is a distance-time graph for a bicycle trip. During which time interval is the cyclist going the fastest?

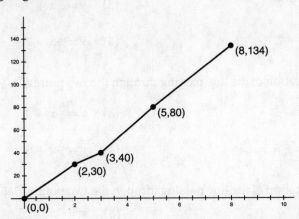

(1)   0 to 2 hours   (3)   3 to 5 hours

(2)   2 to 3 hours   (4)   5 to 8 hours

6. What is the slope of the line defined by the equation $y = -3x + 4$?

(1)   3   (3)   4

(2) -3 (4) -4

7. This is the graph of the solution set of which equation?

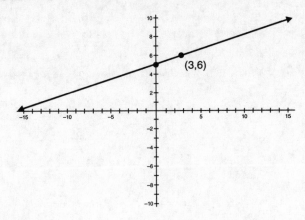

(1) $y = 5x + \dfrac{1}{3}$ (3) $y = 3x + 5$

(2) $y = 5x + 3$ (4) $y = \dfrac{1}{3}x + 5$

8. Find the equation of the line passing through the two points $(0, -7)$ and $(5, 8)$.

   (1) $y = 3x + 7$ (3) $y = \frac{1}{3}x + 7$

   (2) $y = 3x - 7$ (4) $y = \frac{1}{3}x - 7$

9. Find the equation of the line passing through the two points $(4, -2)$ and $(12, 4)$.

   (1) $y = \frac{4}{3}x - 5$ (3) $y = \frac{3}{4}x - 5$

   (2) $y = \frac{4}{3}x + 5$ (4) $y = \frac{3}{4}x + 5$

10. Find the equation of the line passing through the two points $(3, 5)$ and $(3, 8)$.

   (1) $x = -3$ (3) $y = -3$
   (2) $x = 3$ (4) $y = 3$

Solutions for Practice Exercises: Topic 5

1. The x-intercept has a y-coordinate of zero. Substitute zero for y to get
   $2x + 5(0) = 20$, $2x = 20$, $x = 10$. The x-intercept is $(10, 0)$.

   The correct choice is **(1)**.

2. The points on this vertical line all have x-coordinates of 8. The equation is $x = 8$.

   The correct choice is **(1)**.

3. When the four points are plotted on the same graph as the line, only $(6, 8)$ is not on the line.

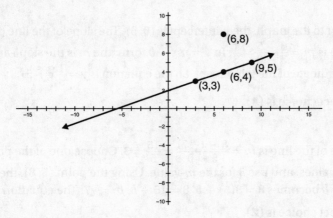

   The correct choice is **(3)**.

4. Using the slope formula $m = \dfrac{y_2 - y_1}{x_2 - x_1}$, with the points $(-2, 1)$ and $(8, 5)$,
   $m = \dfrac{5 - 1}{8 - (-2)} = \dfrac{4}{10} = \dfrac{2}{5}$.

   The correct choice is **(1)**.

5. The interval between 0 and 2 has a slope of $30/2 = 15$. The interval between 2 and 3 has a slope of $10/1 = 10$. The interval between 3 and 5 has a slope of $40/2 = 20$. The interval between 5 and 8 has a slope of $54/3 = 18$. Since the slope represents the speed the bicycle is going, the interval between 3 to 5 hours is the fastest. It can also be seen from the graph that the interval between 3 to 5 hours looks the steepest.

The correct choice is **(3)**.

6. When the equation for a line is in $y = mx + b$ form, m is the slope. For this equation, the coefficient of the x is -3 so the slope of the line is -3.

The correct choice is **(2)**.

7. According to the graph, the y-intercept is $(0, 5)$. The slope of the line through $(0, 5)$ and $(3, 6)$ is $m = \dfrac{6 - 5}{3 - 0} = \dfrac{1}{3}$. In $y = mx + b$ form, the m is the slope and the b is the y-coordinate of the y-intercept. So the equation is $y = \dfrac{1}{3}x + 5$.

The correct choice is **(4)**.

8. The slope of the line is $m = \dfrac{8 - (-7)}{5 - 0} = \dfrac{15}{5} = 3$. Choose one of the points for the x- and y-values, and use 3 for the m-value. Using the point $(5, 8)$, the equation $y = mx + b$ becomes $8 = 3(5) + b$, $8 = 15 + b$, $b = -7$. The equation is $y = 3x - 7$.

The correct choice is **(2)**.

9. The slope of the line is $m = \dfrac{4 - (-2)}{12 - 4} = \dfrac{6}{8} = \dfrac{3}{4}$. Pick one of the points for x and y, and substitute $\dfrac{3}{4}$ for m into $y = mx + b$: $4 = \dfrac{3}{4}(12) + b$, $4 = 9 + b$, $b = -5$. The equation is $y = \dfrac{3}{4}x - 5$.

The correct choice is **(3)**.

10. Since the x-coordinates are the same, the line is a vertical line. Vertical lines have the equation $x = $ constant. Since the x-coordinates are both 3, the equation is $x = 3$.

The correct choice is **(2)**.

5. Graphs of Quadratic Equations

5.1 Graphing a Quadratic Equation with a Chart

A **quadratic equation** can be written in the form $y = ax^2 + bx + c$. The graph of a quadratic equation is always a **parabola**, which resembles the letter U. When the a-value is positive, the parabola looks like a right side up U. When the a-value is negative, the parabola looks like an upside-down U.

- The simplest way to create the graph of a quadratic equation is to choose at least five consecutive x-values and create a chart.

For the equation $y = x^2 - 2x - 3$, the chart could look like this:

x	y
-2	$(-2)^2 - 2(-2) - 3 = 4 + 4 - 3 = 5$
-1	$(-1)^2 - 2(-1) - 3 = 1 + 2 - 3 = 0$
0	$(0)^2 - 2(0) - 3 = -3$
1	$(1)^2 - 2(1) - 3 = 1 - 2 - 3 = -4$
2	$(2)^2 - 2(2) - 3 = 4 - 4 - 3 = -3$

- The graph of these five ordered pairs looks like this:

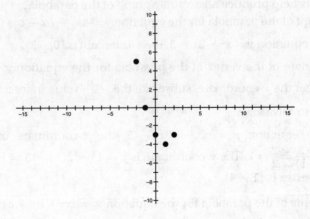

■ *Connect* the points with a U-shaped parabola.

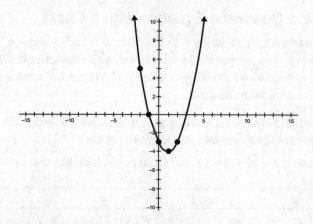

The low point of the parabola is called the **vertex**. In this example, the vertex is (1, −4). The two x-intercepts of the parabola are (−1, 0) and (3, 0). The y-intercept of this parabola is (0, −3).

6.2 Graphing a Parabola by Finding the Vertex and Intercepts

The coordinates of the vertex, the y-intercept, and both x-intercepts can be calculated. These four points help produce an accurate graph of the parabola.

The y-intercept of the parabola for the equation $y = ax^2 + bx + c$ is (0, c).

■ For the equation $y = x^2 - 2x - 3$, the y-intercept is (0, −3).

The x-coordinate of the vertex of the parabola for the equation $y = ax^2 + bx + c$ is $x = -\dfrac{b}{2a}$. To get the y-coordinate, substitute the $-\dfrac{b}{2a}$ value in for x in the equation $y = ax^2 + bx + c$ to solve for y.

■ For the equation $y = x^2 - 2x - 3$, the x-coordinate of the vertex is $x = -\dfrac{-2}{2(1)} = \dfrac{2}{2} = 1$. The y-coordinate is $y = 1^2 - 2(1) - 3 = 1 - 2 - 3 = -4$. So the vertex is (1, −4).

The x-intercepts of the parabola for the equation $y = ax^2 + bx + c$ can be found by solving the quadratic equation $0 = ax^2 + bx + c$.

■ For the equation $y = x^2 - 2x - 3$, this becomes

$$0 = x^2 - 2x - 3$$

This quadratic equation can be solved by factoring.

$$0 = (x + 1)(x - 3)$$
$$x + 1 = 0 \text{ or } x - 3 = 0$$
$$-1 = -1 \quad +3 = +3$$
$$x = -1 \quad \text{or} \quad x = 3$$

The x-intercepts are $(-1, 0)$ and $(3, 0)$.

These four points help make an accurate sketch of the parabola.

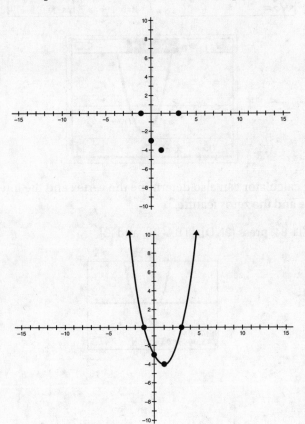

6.3 Graphing Quadratic Equations on the Graphing Calculator

The graphing calculator can easily graph quadratic equations. On the TI-84 press [Y=] enter the equation after "Y1=", and press [ZOOM] and [6]. On the TI-Nspire, from the home screen press [B] for the Graph Scratchpad. Then enter the equation on the entry line and press [enter].

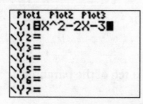
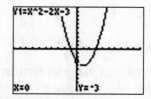

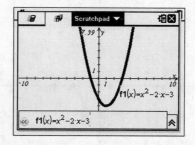

The graphing calculator can also determine the vertex and the intercepts using the min/max feature and the zeros feature.

- For the TI-84, press [2ND], [TRACE], and [3].

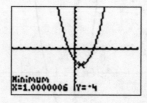

- For the TI-Nspire, press [menu], [6], and [2] to find the minimum point.

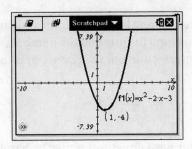

6.4 Using the Graphing Calculator to Solve Quadratic Equations

The solutions to the equation $ax^2 + bx + c = 0$ are also the x-coordinates of the x-intercepts of the parabola defined by $y = ax^2 + bx + c$.

- To solve the equation $x^2 - 2x - 3 = 0$ with the graphing calculator, graph $y = x^2 - 2x - 3$ and then use the zeros feature to find the x-intercepts.

- For the TI-84, enter the equation Y1 $= x^2 - 2x - 3$ and [ZOOM] and [6] to graph. Then press [2ND], [TRACE], and [2] to find each zero.

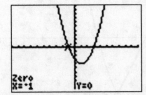 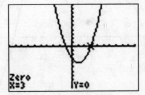

- Here's a new equation to see how the TI-Nspire works. Enter the equation $f1(x) = x^2 - 6x + 8$ to graph it. Then press [menu], [6], and [1] to find each zero.

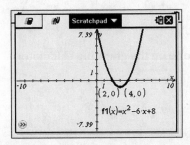

6.5 Solving a Linear-Quadratic System of Equations by Graphing

When a system of equations has one linear equation and one quadratic equation, one way to find the solution is to graph the line and the parabola and find the coordinates of the intersection point, or intersection points. There can be up to two solutions.

For the system of equations

$$y = -2x + 1$$
$$y = x^2 - 2x - 3$$

■ Create the graph for both equations either by hand or with the graphing calculator. The coordinates of the two intersection points are the two solutions to the system of equations. For this system, the two solutions are $x = -2, y = 5$ and $x = 2, y = -3$.

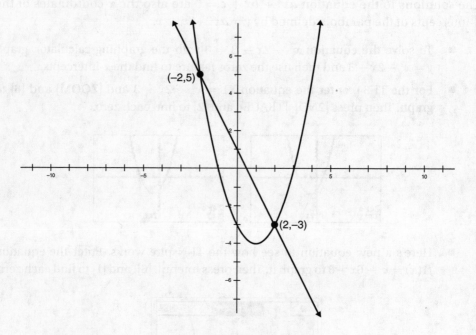

This can also be done on the graphing calculator.

- For the TI-84, graph the two equations and press [2ND], [TRACE], and [5] for each intersection point.

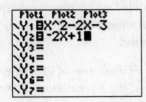
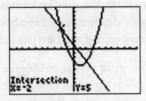
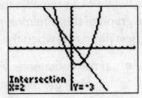

- For the TI-Nspire, graph the two equations and press [menu], [6], and [4] to find the intersection points.

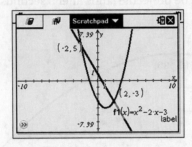

6.6 Word Problems Involving the Graph of a Quadratic Equation

The height of a projectile thrown in the air can be modeled with a quadratic equation. The graph of this equation provides information about when the projectile reaches its highest point and when the projectile lands on the ground.

- If the equation is $h = -16t^2 + 48t + 160$, the y-coordinate of the vertex of the graph is the highest point the projectile reaches. The x-coordinate of the vertex is the amount of time it takes for the projectile to reach its highest point. The x-coordinate of the x-intercept is the amount of time it takes for the projectile to land. For this example, the highest point the projectile reaches is 196 feet high after 1.5 seconds. The projectile lands after 5 seconds.

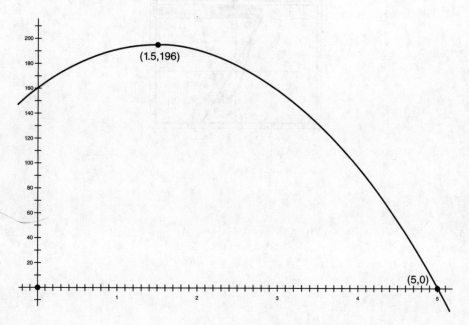

Practice Exercises: Topic 6

1. Which is a point on the graph of the solution set of $y = x^2 + 5x - 2$?

    (1) $(3, 19)$ (3) $(3, 21)$

    (2) $(3, 20)$ (4) $(3, 22)$

2. What are the coordinates of the vertex of the parabola defined by the equation $y = x^2 - 4x - 1$?

    (1) $(-2, 5)$ (3) $(-2, -5)$

    (2) $(2, 5)$ (4) $(2, -5)$

3. $x = -4$ is the x-coordinate of the vertex for the parabola defined by which equation?

    (1) $y = x^2 + 8x + 3$ (3) $y = x^2 + 4x + 3$

    (2) $y = x^2 - 8x + 3$ (4) $y = x^2 - 4x + 3$

4. What could be the equation that determines this parabola?

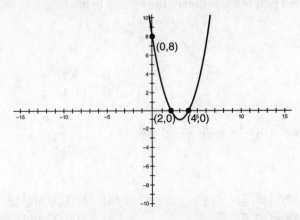

    (1) $y = x^2 - 6x - 8$ (3) $y = x^2 + 6x - 8$

    (2) $y = x^2 - 6x + 8$ (4) $y = x^2 + 6x + 8$

5. Which is the graph of $y = -x^2 + 2x + 3$?

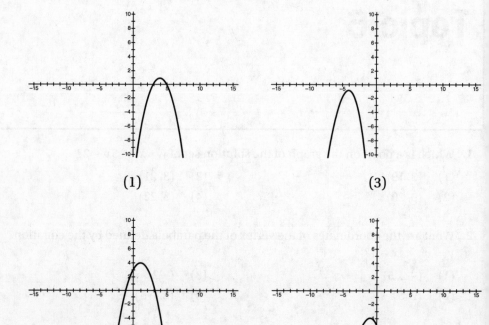

(1)

(3)

(2)

(4)

6. Which ordered pair is a solution to the system below?

$$y = x^2$$
$$y = x + 2$$

(1) $(1, 4)$ (3) $(3, 4)$

(2) $(2, 4)$ (4) $(4, 4)$

7. Solve this system of equations using algebra.

$$y = x^2$$
$$y = 2x + 3$$

(1) $(-1, 1)$ and $(9, 3)$ (3) $(-1, 1)$ and $(-3, 9)$

(2) $(-1, 1)$ and $(3, 9)$ (4) $(1, -1)$ and $(9, 3)$

8. This graph could be used to solve which system of equations?

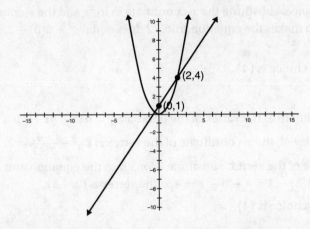

(1) $y = x^2$

$y = \dfrac{2}{3}x + 1$

(2) $y = x^2$

$y = \dfrac{3}{2}x - 1$

(3) $y = x^2$

$y = \dfrac{3}{2}x + 1$

(4) $y = x^2$

$y = \dfrac{2}{3}x - 1$

9. The x-intercepts of the parabola defined by which equation are the solutions to the equation $x^2 + 5x = 15$?

(1) $y = x^2 + 5x + 15$

(2) $y = x^2 + 5x$

(3) $y = x^2 + 5x - 15$

(4) $y = x^2 - 5x - 15$

10. Solve for all values of x, rounded to the nearest hundredth, of $x^2 + 10x + 23 = 0$.

(1) $-3.41, -6.58$

(2) $-3.59, -6.41$

(3) $-3.31, -6.64$

(4) $-3.62, -6.18$

Solutions for Practice Exercises: Topic 6

1. For each choice, substitute the x-coordinate in for x and the y-coordinate in for y to see which makes the equation true. 22 does equal $3^2 + 5(3) - 2$, so the solution is $(3, 22)$.

   The correct choice is **(4)**.

2. The x-coordinate of the vertex of the parabola $y = ax^2 + bx + c$ is $x = \frac{-b}{2a}$. Since a is 1 and b is -4, the x-coordinate of the vertex is $x = -\frac{(-4)}{2(1)} = 2$. To get the y-coordinate of the vertex, substitute 2 for x into the equation and solve for y: $y = 2^2 - 4(2) - 1 = 4 - 8 - 1 = -5$. The vertex is $(2, -5)$.

   The correct choice is **(4)**.

3. Graph each choice on the graphing calculator, and use the minimum feature to find the vertex of each. The vertex of the parabola $y = x^2 + 8x + 3$ is $(-4, -13)$, which has an x-coordinate of -4.

   The correct choice is **(1)**.

4. Since the x-intercepts are $(2, 0)$ and $(4, 0)$, the equation of the parabola is $y = a(x - 2)(x - 4) = a(x^2 - 6x + 8)$. Since the y-intercept is $+8$, $a = 1$ and the equation is $y = x^2 - 6x + 8$.

   The correct choice is **(2)**.

5. The x-coordinate of the vertex must be $x = -\frac{2}{2(-1)} = 1$.

   The correct choice is **(2)**.

6. Substitute x^2 for y in the bottom equation to get $x^2 = x + 2$, $x^2 - x - 2 = 0$, $(x - 2)(x + 1) = 0$, $x = 2$ or $x = -1$. The x-coordinates of the intersection points are 2 and -1. This can also be done with the graphing calculator. The answer is $(2, 4)$.

   The correct choice is **(2)**.

7. Substitute x^2 for y in the bottom equation to get $x^2 = 2x + 3$, $x^2 - 2x - 3 = 0$, $(x - 3)(x + 1) = 0$. So $x = 3$ and $x = -1$ are the x-coordinates of the two intersection points.

    The correct choice is **(2)**.

8. The parabola with vertex $(0, 0)$ passing through $(2, 4)$ is $y = x^2$. The line through $(0, 1)$ and $(2, 4)$ is $y = (3/2)x + 1$.

    The correct choice is **(3)**.

9. This equation can be rewritten as $x^2 + 5x - 15 = 0$. Equations in this form can be solved by finding the x-intercepts of the parabola $y = x^2 + 5x - 15$.

    The correct choice is **(3)**.

10. Using the graphing calculator, graph $y = x^2 + 10x + 23$ and use the "zero" feature to get $x = -3.59$ or $x = -6.41$.

    The correct choice is **(2)**.

7. Linear Inequalities

7.1 One-Variable Linear Inequalities

A linear inequality is like a linear equation, but instead of an = sign there is a $>$, $<$, $\geq$, or $\leq$ sign. Solving a one-variable linear inequality is almost the same as solving a linear equality. The only difference is that when multiplying or dividing both sides by a negative number to eliminate the coefficient, the direction of the inequality sign must be reversed.

$$-2x + 3 < 11$$

- Eliminate the constant 3 by subtracting it from both sides of the inequality.

$$
\begin{aligned}
-2x + 3 &< 11 \\
-3 &= -3 \\
-2x &< 8
\end{aligned}
$$

- Divide both sides by -2 to isolate the x. Because you are dividing by a negative, the direction of the inequality sign must be reversed to keep the equation true. If the -2 were a $+2$, the direction of the inequality sign would not need to be reversed.

$$\frac{-2x}{-2} < \frac{8}{-2}$$

$$x > -4$$

The solution is $x > -4$.

7.2 Graphing Two-Variable Inequalities

$y \leq 2x - 5$ is a two-variable inequality. To graph the two-variable inequality, first graph the line $y = 2x - 5$.

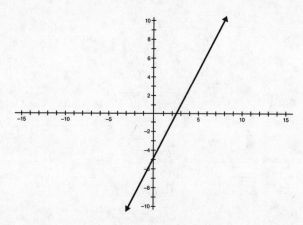

One side of the line needs to be shaded. To determine which side of the line to shade, substitute the ordered pair $(0, 0)$ into the inequality to test if it yields a true inequality.

$$0 \overset{?}{\leq} 2(0) - 5$$
$$0 \overset{?}{\leq} 0 - 5$$
$$0 \leq -5 \text{ is not true.}$$

Since $(0, 0)$ does not satisfy the inequality, shade the side of the line that does not contain $(0, 0)$.

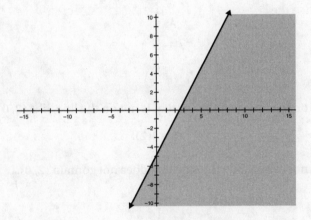

- If the inequality sign is a $<$ or a $>$, the line must be a dotted line. If it is a $\leq$ or a $\geq$, it must be a solid line. If $(0, 0)$ is on the line, an ordered pair that is not on the line must be used to test which side to shade.

- For the graph of $y > 3x$, graph the line $y = 3x$ and make it a dotted line.

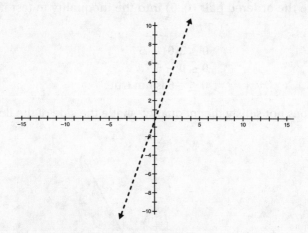

Since $(0, 0)$ is on the line, test a point that is not on the line, like $(2, 0)$.

$$0 > 3(2)$$

Since $0 > 6$ is not true, shade the side that does not contain $(2, 0)$.

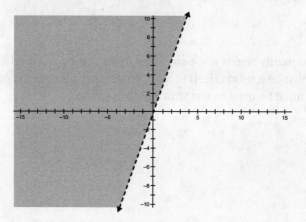

7.3 Graphing Systems of Linear Inequalities

To graph the solution set to a system of linear inequalities, graph both inequalities on the same set of axes and locate the portion of the graph that is shaded twice.

$$y > 2x - 5$$

$$y < -\frac{2}{3}x + 2$$

The portion of the graph that has an S is the solution set. Any point in that region will satisfy the system of inequalities. $(-3, 1)$ is an example of a point that is in this double-shaded region and satisfies both inequalities.

7.4 Graphing Inequalities on the Graphing Calculator

The TI-84 and the TI-Nspire can graph linear inequalities.

- On the TI-84, the shading can be set by moving the cursor to the icon to the left of the Y= and pressing [ENTER] until the icon shows the proper shading.

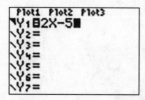 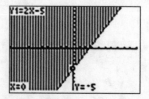

- On the TI-Nspire, delete the "=" from the equation in the entry line and choose the symbol to replace it with.

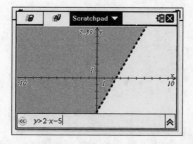

The graphing calculator can also graph systems of linear inequalities by graphing both inequalities on the same set of coordinate axes.

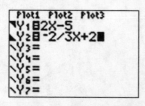 

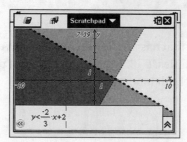

Practice Exercises:
Topic 7

1. What is the solution set for $3x < -18$?

   (1) $x < -6$ (3) $x \geq -6$

   (2) $x > -6$ (4) $x \leq -6$

2. What is the solution set for $-4x \geq 20$?

   (1) $x \geq -5$ (3) $x \leq -5$

   (2) $x > -5$ (4) $x < -5$

3. What is the smallest integer that satisfies the equation $-6x < -18$?

   (1)   3   (3)   5

   (2)   4   (4)   6

4. Which is the graph of $y < x + 4$?

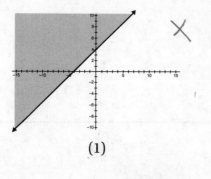 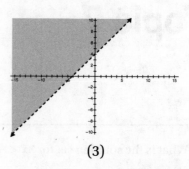

(1)

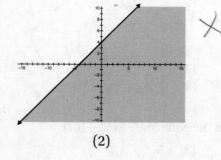 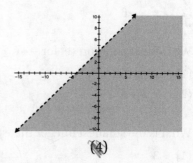

(2)

(3)

(4)

5. Below is the graph of $x + y \leq 8$. Which is a point in the solution set?

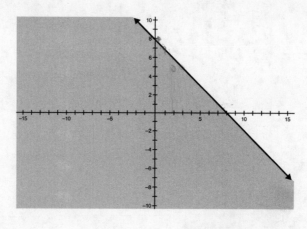

(1)　　(1, 8)

(2)　　(2, 7)

(3)　　(4, 5)

(4)　　(3, 5)

6. All of these ordered pairs are part of the shaded region for the graph of $2x + y \leq 12$ except

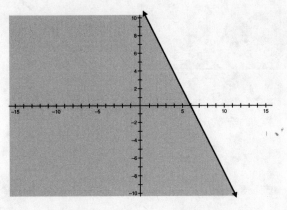

(1) $(2, 8)$ (3) $(5, 3)$

(2) $(3, 6)$ (4) $(4, 4)$

7. Which graph shows the solution to the system of inequalities?

$$y < 2x + 1$$
$$y > \frac{1}{3}x + 4$$

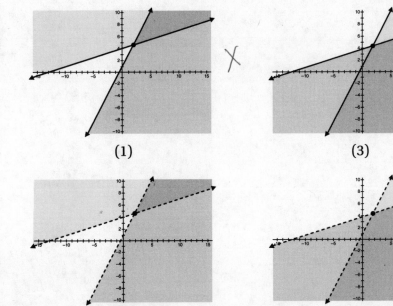

(1)   (3)

(2)   (4)

8. Which system of inequalities does the following graph show the solution for?

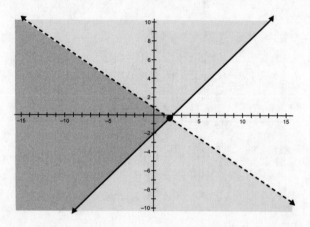

(1) $y \geq x - 2$ ✗

$y < -\frac{2}{3}x + 1$

(2) $y \leq x - 2$

$y > -\frac{2}{3}x + 1$

(3) $y < x - 2$ ✗

$y \geq -\frac{2}{3}x + 1$

(4) $y > x - 2$

$y \leq -\frac{2}{3}x + 1$

9. Which graph has the solution set shaded in for the following system of inequalities?

$$y \leq -x + 6$$
$$y \geq \frac{1}{2}x - 1$$

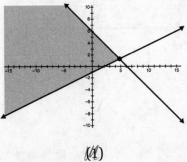

(1)

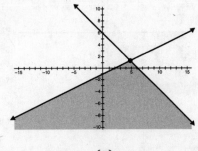

(3)

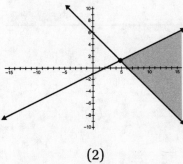

(2)

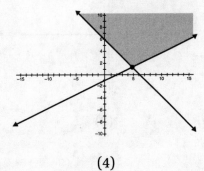

(4)

10. Which is the graph of the following system of inequalities?

$$y \geq 0$$
$$x \leq 0$$

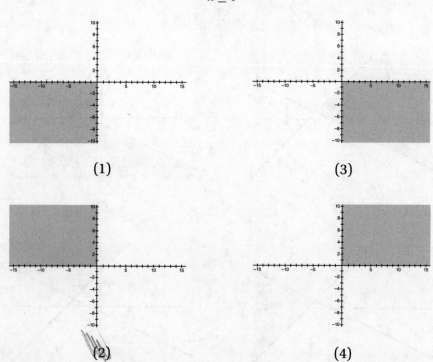

(1)

(3)

(2)

(4)

Solutions for Practice Exercises: Topic 7

1. Divide both sides of the inequality by 3 to get $x < -6$.

   The correct choice is **(1)**.

2. Divide both sides of the inequality by -4. Switch the direction of the inequality sign because you divided by a negative to get $x \leq -5$.

   The correct choice is **(3)**.

3. Divide both sides of the inequality by -6 and switch the direction of the inequality sign to get $x > 3$. The smallest integer that satisfies the equation is 4. The answer is not 3 since it is not true that $3 > 3$.

   The correct choice is **(2)**.

4. Because it is a $<$ and not a $\leq$ sign, the line must be dotted. Test to see if $(0, 0)$ is in the solution set by checking if $0 < 0 + 4$ is true. Since it is true, the side of the line containing $(0, 0)$ must be shaded.

   The correct choice is **(4)**.

5. When all four choices are plotted on the graph, three of them are not in the shaded region. The point $(3, 5)$ is on the line, but since there is a $\leq$ sign, the line is part of the solution set.

   The correct choice is **(4)**.

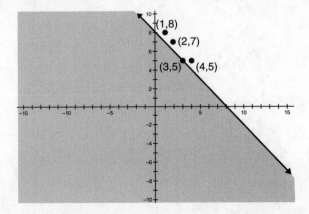

6. Three of the points are on the line, which is part of the solution set because it is a
≤ sign. The point (5, 3) is not in the shaded area.

The correct choice is **(3)**.

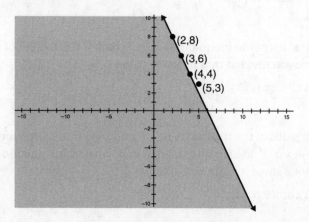

7. Both of the lines have to be dotted because of the inequality signs < and >. This
eliminates choices (1) and (3). Both choices (2) and (4) have the region below
the line $y = 2x + 1$ shaded. The difference between choices (2) and (4) is that
choice (2) has the region above $y = (1/3)x + 4$ shaded, and choice (4) has the
region below $y = (1/3)x + 4$ shaded. To check which side is correct, substitute
(0, 0) into the equation $y > (1/3)x + 4$ to get $0 > (1/3)(0) + 4$. Since 0 is not
greater than 4, (0, 0) is not part of the solution set to $y > (1/3)x + 4$. The side of
the line $y = (1/3)x + 4$ that contains (0, 0) should not be shaded, so the region
above the line $y = (1/3)x + 4$ should be shaded.

The correct choice is **(2)**.

8. Since the line $y = x - 2$ is solid, choices (3) and (4) can be eliminated. Substituting $(0, 0)$ into both inequalities in choice (1) makes them both true, while substituting $(0, 0)$ into both inequalities in choice (2) makes them both false.

The correct choice is **(1)**.

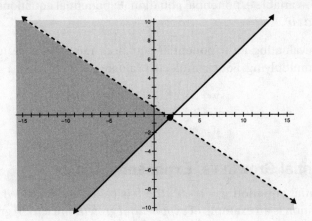

9. To check choice (1), pick a point in the shaded region for that choice and see if it satisfies both inequalities. Since $(0, 0)$ is a point in the shaded region, check to see if $0 \leq -0 + 6$ and $0 \geq (1/2) \cdot 0 - 1$. The point $(0, 0)$ does make both inequalities true.

The correct choice is **(1)**.

10. The points on or above the x-axis make $y \geq 0$ true. The points on or to the left of the y-axis make $x \leq 0$ true. The points that satisfy both inequalities must be above the x-axis and to the left of the y-axis.

The correct choice is **(2)**.

8. Exponential Equations

8.1 Evaluating Exponential Expressions

An **exponential equation** is one where one of the variables is an exponent. The equation $y = 3 \cdot 2^x$ is a two-variable exponential equation. Exponential equations can be written in the form $y = a \cdot b^x$.

- When evaluating an exponential equation, raise the base to the exponent before multiplying. For example, to evaluate $y = 3 \cdot 2^x$ when $x = 4$, it becomes

$$y = 3 \cdot 2^4$$
$$y = 3 \cdot 16 \ (\text{NOT } y = 6^4)$$
$$y = 48$$

8.2 Exponential Growth vs. Exponential Decay

In the exponential equation $y = a \cdot b^x$, the b is called the **base**. When b is greater than 1, the equation is an example of *exponential growth* since $a \cdot b^x$ grows as x grows. When b is between 0 and 1, the equation is an example of *exponential decay* since $a \cdot b^x$ gets smaller as x grows.

$y = 3 \cdot 2^x$ is an example of exponential growth since $b = 2$, which is greater than 1.

$y = 3 \cdot (0.7)^x$ is an example of exponential decay since $b = 0.7$, which is between 0 and 1.

8.3 Graphs of Exponential Equations

A table of values for a two-variable exponential equation, like $y = 3 \cdot 2^x$, looks like this:

x	y
-3	$3 \cdot 2^{-3} = 3 \cdot \dfrac{1}{2^3} = 3 \cdot \dfrac{1}{8} = \dfrac{3}{8}$
-2	$3 \cdot 2^{-2} = 3 \cdot \dfrac{1}{2^2} = 3 \cdot \dfrac{1}{4} = \dfrac{3}{4}$
-1	$3 \cdot 2^{-1} = 3 \cdot \dfrac{1}{2^1} = 3 \cdot \dfrac{1}{2} = \dfrac{3}{2}$
0	$3 \cdot 2^0 = 3 \cdot 1 = 3$
1	$3 \cdot 2^1 = 3 \cdot 2 = 6$
2	$3 \cdot 2^2 = 3 \cdot 4 = 12$
3	$3 \cdot 2^3 = 3 \cdot 8 = 24$

Remember that raising a number to a negative power, like a^{-x}, is equivalent to $\frac{1}{a^x}$. For example, $2^{-3} = \frac{1}{2^3} = \frac{1}{8}$.

The graph for this equation has this shape:

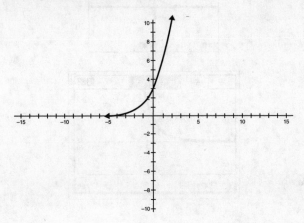

For exponential decay, like $y = 3 \cdot \left(\frac{1}{2}\right)^x$, the graph has this shape:

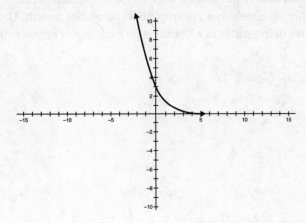

Exponential equations can also be graphed on the graphing calculator.

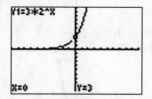

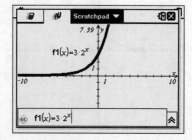

8.4 Real-World Scenarios Involving Exponential Equations

Many real-world scenarios can be modeled with exponential equations. The population of a country over time is generally an example of exponential growth. The temperature of food over time after being put into a freezer is an example of exponential decay.

Practice Exercises: Topic 8

1. If $x = 2$ and $y = 3^x$, solve for y.

   (1)   8
   (2)   9

   (3)   10
   (4)   11

2. If $x = 3$ and $y = 2 \cdot 3^x$, solve for y.

   (1)   216
   (2)   27

   (3)   54
   (4)   18

3. Which ordered pair is in the solution set of $y = 5 \cdot 2^x$?

   (1) $(0, 0)$
   (2) $(2, 25)$

   (3) $(3, 40)$
   (4) $(3, 100)$

4. Which is the graph of $y = 2^x$?

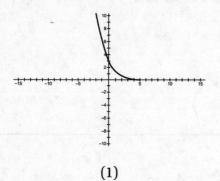

(1)

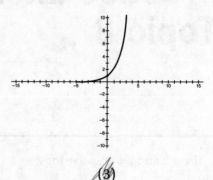

(3)

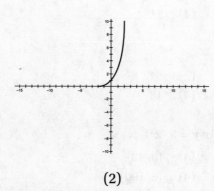

(2)

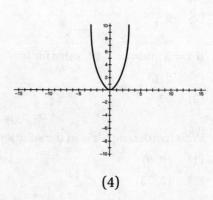

(4)

5. Below is the graph of which equation?

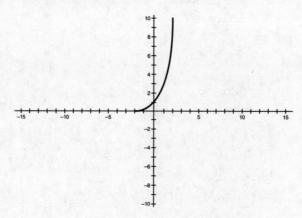

(1) $y = \left(\frac{1}{3}\right)^x$

(2) $y = 10^x$

(3) $y = 4^x$

(4) $y = 3^x$

6. When is the graph of $y = 1.5^x$ increasing?

(1) Always

(2) Never

(3) When $x \geq 0$

(4) When $x \leq 0$

7. Below is the graph of $y = b^x$. What is true about the value of b?

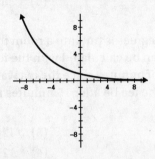

(1) b must be greater than 1.

(2) b must be less than 1.

(3) b must be less than 0.

(4) b must be less than -1.

8. What type of equation has a graph like the one below?

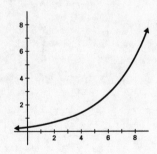

(1) Linear

(2) Exponential

(3) Quadratic

(4) None of the above

9. The population of a country can be modeled with the equation $P = 250 \cdot 1.07^t$, where P is the population in millions and t is the number of years since 2020. According to this model, rounded to the nearest ten million, what will the population of this country be in 2029?

   (1) 450,000,000

   (2) 460,000,000

   (3) 470,000,000

   (4) 480,000,000

10. A cup of tea that is 200 degrees is put into a room that is 80 degrees. The temperature of the tea can be calculated with the formula $t = 200 \cdot 0.9^m + 80$, where m is the number of minutes since the tea was put into the room. What will the temperature of the tea be after 15 minutes rounded to the nearest degree?

    (1) 116 degrees

    (2) 118 degrees

    (3) 120 degrees

    (4) 121 degrees

Solutions for Practice Exercises: Topic 8

1. Substitute 2 for x. The equation becomes $y = 3^2 = 9$.

    The correct choice is **(2)**.

2. Substitute 3 for x. The equation becomes $y = 2 \cdot 3^3 = 2 \cdot 27 = 54$.

    The correct choice is **(3)**.

3. Substitute each ordered pair into the equation to see which makes it true. For the ordered pair $(3, 40)$, $5 \cdot 2^3 = 5 \cdot 8 = 40$.

    The correct choice is **(3)**.

4. The graph for $y = 2^x$ must pass through the points $(0, 1)$, $(1, 2)$, and $(2, 4)$. It also passes through $(-1, 1/2)$ and $(-2, 1/4)$.

    The correct choice is **(3)**.

5. Since this graph passes through $(0, 1)$, $(1, 3)$, and $(2, 9)$, it is the equation $y = 3^x$.

    The correct choice is **(4)**.

6. An exponential graph with a base greater than 1 is always increasing.

    The correct choice is **(1)**.

7. When an exponential graph is decreasing, the base is between 0 and 1.

    The correct choice is **(2)**.

8. This graph has the standard shape of an exponential curve based on an equation $y = b^x$ with $b > 1$. An exponential curve like this starts off relatively flat for x-values close to zero and then increases rapidly.

    The correct choice is **(2)**.

9. Since 2029 is 9 years past 2020, substitute 9 for t to get $P = 250 \cdot 1.07^9 = 459.6$. In millions, this is approximately 460,000,000.

   The correct choice is **(2)**.

10. Substitute 15 for m to get $t = 200 \cdot 0.9^{15} + 80 = 121.1782$ or 121.2, which rounds to 121.

    The correct choice is **(4)**.

9. Creating and Interpreting Equations

9.1 Creating and Interpreting Linear Equations

Many real-world situations can be modeled with a linear equation in the form $y = mx + b$. The m and the b have to be replaced with the appropriate values from the situation. In general, the b-value is the starting value and the m-value is the amount the total changes each time the x-variable increases.

■ If a carnival costs \$10 admission and \$3 for each ride, the 10 and the 3 can be used in a linear equation. Since the 10 is the starting amount, it would take the place of the b in the equation. Since 3 is the amount the total increases by for each new ride, it would take the place of m in the equation.

The equation could be written as $y = 3x + 10$, where y is the total cost and x is the number of rides. Instead of x and y, the total could be represented by the variable T, whereas the number of rides could be represented by the variable R to form the equation $T = 3R + 10$.

When given an equation modeling a real-world situation, it is also possible to interpret what the values for m and b represent. The b represents the starting value, and the m represents the amount that the total changes for each increase in x.

■ If the equation for the cost of a pizza with N toppings is $C = 2N + 12$, you could be asked to interpret what the 2 and the 12 represent. Since the 2 is in the place of the m in $y = mx + b$, it represents the cost of each topping. Since the 12 is in the place of the b in $y = mx + b$, it represents the cost of the pizza before any toppings are added.

9.2 Creating and Interpreting Exponential Equations

An **exponential equation** is a good model for many real-world situations, including population growth, compound interest, and liquid cooling. Exponential equations have the form $y = a \cdot (1 + r)^x$, where the a represents the starting value and the r represents the growth rate.

■ If the population of a town is 10,000 people and the annual growth rate is 7%, then the equation that relates total population, P, to the number of years that have passed, T, is $P = 10{,}000 \cdot 1.07^T$. Since 10,000 is the starting value, it takes the place of the a, and since the growth rate is 0.07, it takes the place of the r in the equation $y = a \cdot (1 + r)^x$.

If an exponential equation that models a real-world situation is given, it is possible to interpret what the numbers in the equation represent.

- If, in another town, the equation relating its population to the number of years that have passed is $P = 20{,}000 \cdot 1.09^T$, you could be asked to interpret what the numbers 20,000 and 0.09 represent. In this case, the 20,000 represents the starting population and the 0.09 represents the growth rate.

- When a ball is dropped, each bounce is 80% as high as the previous bounce. If the ball is dropped from a window 50 feet above the ground, the equation that relates the height of the bounce (H) to the number of bounces (B) will be $H = 50 \cdot 0.80^B$. Since the starting height is 50, it takes the place of the a, and since 0.80 can be expressed as $(1 - 0.20)$, the r-value is -0.20, which is the growth rate. A -0.20 growth rate can also be called a *decay* rate of 0.20.

Practice Exercises: Topic 9

1. It costs $10 to go to the movies and $3 for each bag of popcorn. Which equation relates the total cost (C) to the number of bags of popcorn purchased (P)?

(1) $C = 3P + 10$ (3) $P = 3C + 10$

(2) $C = 10P + 3$ (4) $P = 10C + 3$

2. A tablet costs $400 and $2 for each app. Which equation relates the total cost (C) to the number of apps purchased (A)?

(1) $A = 2 + 400C$ (3) $C = 2 + 400A$

(2) $A = 400 + 2C$ (4) $C = 400 + 2A$

3. A cable TV plan costs $80 a month plus $10 extra for each premium channel. Which equation relates the monthly bill (B) to the number of premium channels ordered (C)?

(1) $B = 80C + 10$ (3) $C = 80B + 10$

(2) $B = 10C + 80$ (4) $C = 10B + 80$

4. Lydia wants to buy an ice cream cone and add some toppings. The equation that relates the total cost for the ice cream cone and N toppings is $P = 1.50N + 5$. What does the number 5 in the equation represent?

(1) The cost of the ice cream cone

(2) The cost of each topping

(3) The cost of all N toppings

(4) The total cost of the ice cream cone and all N toppings

5. Amelia buys an empty sticker album and some sticker sheets. The equation that relates the total cost for the empty sticker album and N sticker sheets is $P = 0.75N + 3.00$. What does the number 0.75 in the equation represent?

   (1)   The cost of the empty sticker album

   (2)   The cost of each sticker sheet

   (3)   The cost of all N sticker sheets

   (4)   The total cost of the empty sticker album and all N sticker sheets

6. There were 900 birds in a forest. Each year the bird population increases by 12%. Which equation relates the bird population (P) to the number of years that have passed?

   (1) $P = 900(1.12)^t$ (3) $P = 900(0.88)^t$

   (2) $P = 900(0.12)^t$ (4) $t = 900(1.12)^P$

7. A bouncing ball is dropped from 20 feet high. After each bounce, the height of the next bounce is 65% as high as the last bounce. Which equation relates the height of the bounce (H) to the number of bounces that have happened (N)?

   (1) $H = 20(0.35)^N$ (3) $H = 20(0.65)^N$

   (2) $H = 20(1.65)^N$ (4) $N = 20(0.65)^H$

8. The population (P) of a town after t years can be modeled with the equation $P = 20,000(1.07)^t$. What does the 20,000 represent?

   (1)   The growth rate

   (2)   The percent increase each year

   (3)   The population after t years

   (4)   The starting population of the town

9. After Allie takes some medicine, the number of milligrams of medicine (M) remaining in her body after t minutes can be modeled with the equation $M = 200(1 - 0.27)^t$. Which number represents the decay rate?

(1) 0.27   (3) 0.73

(2) 200   (4) 146

10. Mason puts money into a bank that offers interest compounded annually. The formula relating the amount of money in the bank (A) to the number of years it has been in the bank (t) is $A = 800(1.2)^t$. What is the interest rate the bank offers?

(1) 1.2%   (3) 20%

(2) 2%   (4) 120%

Solutions for Practice Exercises: Topic 9

1. P bags of popcorn cost \$3P. The entrance price is \$10. So the total price is $3P + 10$.

   The correct choice is **(1)**.

2. A apps cost \$2A. The tablet costs \$400. So the total price is $2A + 400$ or $400 + 2A$.

   The correct choice is **(4)**.

3. C premium channels cost \$10C. The monthly cost is \$80. So the total price is $10C + 80$.

   The correct choice is **(2)**.

4. The 1.50 is the cost for each topping, and the 5 is the price of the ice cream cone. In general, the constant represents the fixed cost.

   The correct choice is **(1)**.

5. The 3.00 is the part that does not depend on the value of N. The $0.75N$ increases by 0.75 each time N increases by 1, so the 0.75 represents the cost of each sticker sheet.

   The correct choice is **(2)**.

6. In an exponential equation of the form $P = a \cdot (1 + r)^t$, the r-value is the percent increase (or decrease if r is negative) and the a is the initial value. When $r = 0.12$ and $a = 900$, the equation is $P = 900(1.12)^t$.

   The correct choice is **(1)**.

7. The height of each bounce is 0.65 multiplied by the height of the previous bounce. The initial height is 20 feet. After one bounce, the second is $20 \cdot 0.65$. After two bounces, the ball has a height of $20 \cdot 0.65 \cdot 0.65$. In general, after N bounces, the height is $H = 20 \cdot 0.65^N$.

   The correct choice is **(3)**.

8. In an equation of the form $P = a \cdot (1 + r)^t$, the r is the percent increase each year and the a is the initial value at $t = 0$. The 20,000 therefore represents the starting population of the town.

The correct choice is **(4)**.

9. In an equation of the form $M = a \cdot (1 - r)^t$, the r is the decay rate. Since the $(1 - 0.27)$ is being raised to the t power in this equation, the decay rate is 0.27.

The correct choice is **(1)**.

10. The interest rate is the percent increase each year. The percent increase in an equation of the form $A = P \cdot (1 + r)^t$ is the r-variable. Since $1.2 = 1 + 0.2$, the r-value is 0.2, which is 20%.

The correct choice is **(3)**.

10. Functions

10.1 Different Representations of Functions

A **function** is like a machine that takes a number as an input and outputs a number. Functions are often named with lowercase letters like f and g.

- If a function f takes the number 2 as an input and outputs the number 7, we say $f(2) = 7$. The number in the parentheses is the number that is input into the function. The number after the equal sign is the number that is output from the function.

A function can be represented in several different ways. Here are the most common ways:

1. As a list of ordered pairs

   If the function f is defined as $f = \{(1, 4), (2, 7), (3, 10), (4, 13), (5, 16)\}$, the numbers in the parentheses represent an input value and an output value for the function. In this example, the point $(1, 4)$ in the definition means that if 1 is put into the function, 4 is output from the function, or $f(1) = 4$. Likewise, $f(2) = 7$, $f(3) = 10, f(4) = 13$, and $f(5) = 16$.

2. As an equation

   A function can be defined by an equation that allows you to calculate the output value for a given input value. An example is the definition $f(x) = 3x + 1$. With this definition, it is possible to calculate the output value for any input value. For example, to calculate $f(10)$, substitute the number 10 into the equation $f(x) = 3x + 1$ to become $f(10) = 3(10) + 1 = 30 + 1 = 31$, so $f(10) = 31$.

3. As a graph

   - If the function is defined as a graph, determine the value of $f(2)$ by finding a point on the graph that has an x-coordinate of 2. The y-coordinate of that point is the value of $f(2)$. Since the point with an x-coordinate of 2 is $(2, 7)$, the value of $f(2) = 7$.

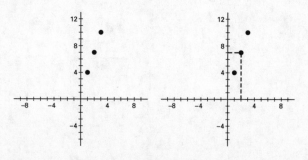

10.2 Domain and Range of Functions

The possible input values of a function are called the *domain* of the function. The possible output values are called the *range* of the function.

For the function $f = \{(1, 4), (2, 7), (3, 10), (4, 13), (5, 16)\}$, the domain is the set of input values $\{1, 2, 3, 4, 5\}$. The range is the set of output values $\{4, 7, 10, 13, 16\}$.

- When a function is described as a graph, the domain is the set of x-coordinates of all the points on the graph and the range is the set of y-coordinates of all the points on the graph.

For this graph, the domain is $\{1, 2, 3, 4\}$ and the range is $\{3, 5, 8\}$.

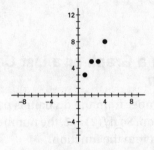

For this graph, the domain is $2 \leq x \leq 5$ and the range is $3 \leq y \leq 7$.

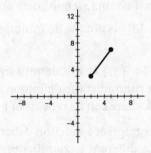

In a real-world situation, the domain is often a special subset of the real numbers. If the function has as its input the number of cars a salesperson sells in a month, the domain would be the set of nonnegative integers $\{0, 1, 2, \ldots\}$ since fractions of a car or negative cars cannot be sold.

10.3 Graphing Functions

The graph of the function $f(x) = x^2$ is the same as the graph of $y = x^2$. All the methods for graphing by hand or with the graphing calculator described earlier can be used for functions.

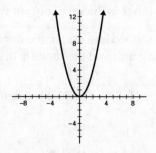

10.4 How to Tell When a Graph or a List Cannot Define a Function

In a function, each time a number from the domain is put into the function, the same value is output from the function. So if $f(2) = 7$, the number 7 will always be output from the function whenever 2 is put into the function.

■   When a function is defined as a list of ordered pairs, then there will never be two ordered pairs with the same x-coordinate but different y-coordinates.

$f = \{(1, 4), (2, 7), (2, 10), (3, 13)\}$ is not the definition of a function since $f(2)$ can be 7 or 10.

$g = \{(1, 4), (2, 7), (3, 7), (4, 13)\}$ is the definition of a function. The fact that $g(2) = 7$ and $g(3) = 7$ does not contradict the definition of a function. As long as there are no repeats of x-coordinates, there can be repeats of y-coordinates.

■   A graph will not be the graph of a function if there are two points that have the same x-coordinate but different y-coordinates. Graphs that cannot be functions fail the **vertical line test**, which means that at least one vertical line will pass through two or more points. All the points a vertical line passes through will have the same x-coordinates but different y-coordinates.

These graphs cannot be graphs of functions since they fail the vertical line test at at least one location.

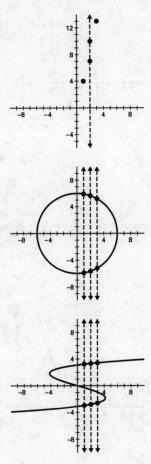

These graphs can be graphs of functions since they pass the vertical line test at all locations.

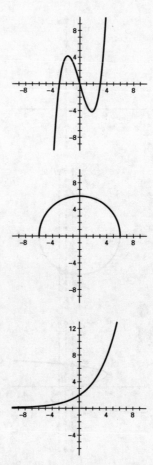

10.5 Graphing Transformed Functions

- If the graph of $y = f(x)$ is already known, the graph of $y = f(x+a)$, $y = f(x-a)$, $y = f(x) + a$, and $y = f(x) - a$ can be easily graphed by knowing the four basic transformations.

- If the graph of $y = f(x)$ looks like this:

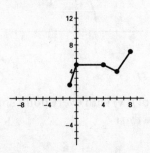

then the graph of

1. $f(x) + a$ will be the graph of $f(x)$ with every point shifted *up* by a units. The graph of $f(x) + 2$ looks like this:

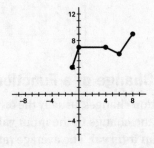

2. $f(x) - a$ will be the graph of $f(x)$ with every point shifted *down* by a units. The graph of $f(x) - 2$ looks like this:

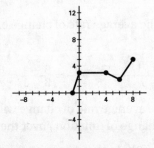

3. $f(x + a)$ will be the graph of $f(x)$ with every point shifted *left* by a units. The graph of $f(x + 2)$ looks like this:

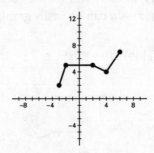

4. $f(x - a)$ will be the graph of $f(x)$ with every point shifted *right* by a units. The graph of $f(x - 2)$ looks like this:

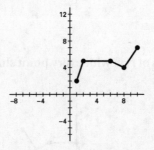

10.6 Average Rate of Change of a Function Over an Interval

When the input value of a function changes, usually the output also changes. The change in the output value related to the change in the input value is called the **average rate of change of a function over an interval**. The average rate of change of function f over the interval from a to b can be calculated by the following formula:

$$\frac{f(b) - f(a)}{b - a}$$

If the function is $f(x) = x^2$, the average rate of change of function f over the interval 1 to 3 is equal to:

$$\frac{f(3) - f(1)}{3 - 1} = \frac{3^2 - 1^2}{3 - 1} = \frac{9 - 1}{3 - 1} = \frac{8}{2} = 4$$

Over a different interval, the average rate of change will usually be different. For this function, the average rate of change of function f over the interval 2 to 4 is equal to:

$$\frac{f(4) - f(2)}{4 - 2} = \frac{4^2 - 2^2}{4 - 2} = \frac{16 - 4}{4 - 2} = \frac{12}{2} = 6$$

Practice Exercises: Topic 10

1. If a function f is defined as $f = \{(1, 2), (2, 3), (3, 1), (4, 4)\}$, what is $f(2)$?

   (1) 1   (3) 3

   (2) 2   (4) 4

2. Which of the following *cannot* be the definition of a function?

   (1) $f = \{(1, 5), (2, 7), (2, 8), (4, 9)\}$

   (2) $f = \{(1, 2), (2, 2), (3, 2), (4, 2)\}$

   (3) $f = \{(0, 0), (1, 1), (-1, 1), (2, 4), (-2, 4)\}$

   (4) $f = \{(6, 1)\}$

3. What is the domain of the function defined as $f = \{(1, 4), (3, 7), (4, 8), (5, 8)\}$?

   (1) $\{4, 7, 8\}$ (3) $\{1, 3, 4, 5\}$

   (2) $\{1, 3, 4, 5, 7, 8\}$ (4) $\{4\}$

4. Below is the graph of $y = f(x)$. What is the value of $f(3)$?

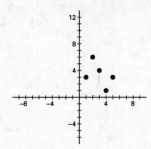

   (1) 1   (3) 3

   (2) 2   (4) 4

5. Which is the graph of a function?

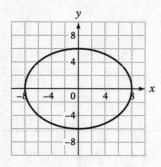

(1)

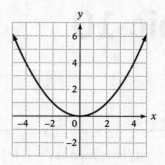

(3)

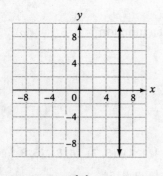

(2)

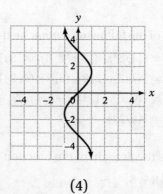

(4)

6. If $g(x) = -x^2 + 7x + 1$, what is $g(2)$?

(1)   11   (3)   27

(2)   19   (4)   35

7. If below is the graph of $y = f(x)$, which is the graph of $y = f(x) - 5$?

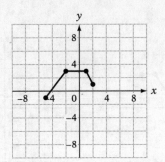

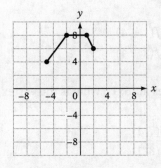

(1)

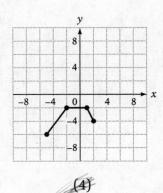

(3)

(2)

(4)

8. What is the average rate of change of the function $f(x) = x^2 + 5x$ over the interval 1 to 4?

(1)   1

(2)   3

(3)   10

(4)   30

9. Below is the graph of $f(x)$ on the left and $g(x)$ on the right. Which is equivalent to $g(x)$?

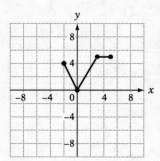 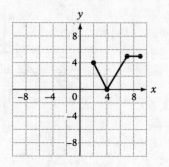

(1) $f(x) + 4$ (3) $f(x + 4)$

(2) $f(x) - 4$ (4) $f(x - 4)$

10. If the graph of $y = f(x)$ is a parabola with the vertex at $(5, 1)$, what is the vertex of the graph of the parabola $y = f(x - 2)$?

(1) $(5, 3)$ (3) $(7, 1)$

(2) $(5, -1)$ (4) $(3, 1)$

◆olutions for Practice Exercises: Topic 10

1. Since the ordered pair $(2, 3)$ is in the set, $f(2) = 3$.

   The correct choice is **(3)**.

2. A set of ordered pairs cannot be the definition of a function if there are two ordered pairs with the same x-coordinate but different y-coordinates. In choice (1), there are two ordered pairs with x-coordinates of 2, $(2, 7)$ and $(2, 8)$. So that cannot be the definition of a function.

   The correct choice is **(1)**.

3. The domain is the set of x-coordinates. In this example, the x-coordinates are 1, 3, 4, and 5.

   The correct choice is **(3)**.

4. There is a point at $(3, 4)$, so $f(3) = 4$.

   The correct choice is **(4)**.

5. Choices (1), (2), and (4) all fail the vertical line test since at least one vertical line can pass through more than one point on them. Choice (3) can be the graph of a function since there is no possible vertical line that would pass through more than one point on it.

   The correct choice is **(3)**.

6. Substitute 2 for x to get $g(2) = -2^2 + 7 \cdot 2 + 1 = -4 + 14 + 1 = 11$.

   Note that $-2^2 = -4$ and not 4. If there were parentheses around -2, as in $(-2)^2$, then it would equal 4.

   The correct choice is **(1)**.

7. The graph of $y = f(x) - 5$ is the same as the graph of $y = f(x)$ with each point shifted 5 units down.

   The correct choice is **(4)**.

8. The average rate of change of a function over the interval a to b is $\dfrac{f(b) - f(a)}{b - a}$. For this question:

$$\frac{f(4) - f(1)}{4 - 1} = \frac{(4^2 + 5 \cdot 4) - (1^2 + 5 \cdot 1)}{4 - 1} = \frac{36 - 6}{4 - 1} = \frac{30}{3} = 10$$

The correct choice is **(3)**.

9. The graph of $g(x)$ is the same as the graph of $y = f(x)$ with each point shifted 4 units to the right. $g(x)$ must be equivalent to $f(x - 4)$.

The correct choice is **(4)**.

10. The graph of $f(x - 2)$ is the same as the graph of $f(x)$ with each point shifted 2 units to the right. If the vertex of the parabola for $f(x)$ is at $(5, 1)$, the vertex of the parabola for $f(x - 2)$ is 2 units to the right of $(5, 1)$, which is at $(7, 1)$.

The correct choice is **(3)**.

1. Sequences

1.1 Types of Sequences

sequence is a series of numbers that can be predicted by some kind of pattern. Two f the most common types of sequences are **arithmetic sequences** and **geometric** :quences.

- The sequence 2, 5, 8, 11, 14, ... is an example of an arithmetic sequence since each term after the first can be obtained by adding the same number, 3, to the previous term.

- The sequence 2, 6, 18, 54, 162, ... is an example of a geometric sequence since each term after the first can be obtained by multiplying the same number, 3, by the previous term.

The terms in the sequence can be described with subscript notation. If the first terms f sequence a are 2, 5, 8, 11, and 14, then $a_1 = 2$, $a_2 = 5$, $a_3 = 8$, $a_4 = 11$, and $a_5 = 14$.

1.2 Describing a Sequence with a Direct Formula

he terms of the sequence 2, 5, 8, 11, ... can be described by the formula $_n = 2 + 3(n - 1)$ or $a(n) = 2 + 3(n - 1)$. If you substitute $n = 1$ into the formula, it :comes $a_1 = 2 + 3(1 - 1) = 2 + 3(0) = 2 + 0 = 2$.

- For any arithmetic sequence, the direct formula for the nth term is $a_n = a_1 + d(n - 1)$ where a_1 is the first term of the sequence and d is the common difference between two consecutive terms. For the sequence 2, 5, 8, 11, ..., $a_1 = 2$ and $d = 3$.

- For any geometric sequence, the direct formula for the nth term is $a_n = a_1 \cdot r^{n-1}$, where a_1 is the first term of the sequence and r is the common ratio between two consecutive terms. For the sequence 2, 6, 18, 54, ..., $a_1 = 2$ and $r = 3$ since if you divide any term by the previous term you get 3.

Practice Exercises: Topic 11

1. What type of sequence is 3, 7, 11, 15, ...?
   - (1) Increasing arithmetic
   - (2) Decreasing arithmetic
   - (3) Increasing geometric
   - (4) Decreasing geometric

2. What type of sequence is 4, 8, 16, 32, ...?
   - (1) Increasing arithmetic
   - (2) Decreasing arithmetic
   - (3) Increasing geometric
   - (4) Decreasing geometric

3. What is the next number in the sequence $20, 10, 5, \frac{5}{2}$?
   - (1) $\frac{5}{4}$
   - (2) $\frac{5}{8}$
   - (3) $\frac{5}{16}$
   - (4) $\frac{5}{32}$

4. If in an arithmetic sequence, $a_4 = 17$ and $a_8 = 33$, what is the value of a_5?
   - (1) 4
   - (2) 20
   - (3) 21
   - (4) 22

5. If in an arithmetic sequence, $a_2 = 36$ and $a_5 = 21$, what is the value of a_7?
   - (1) −5
   - (2) 5
   - (3) 11
   - (4) 31

6. If in an arithmetic sequence, $a_{20} = 122$ and $a_{45} = 272$, what is the value of a_1?

   (1) 8   (3) 7

   (2) 6   (4) 5

7. If in an arithmetic sequence, $a_{58} = 395$ and $a_{74} = 507$, what is the value of a_{211}?

   (1) 1,460   (3) 1,466

   (2) 1,463   (4) 1,469

   [handwritten: $a_{74} = a_1 + 7(74-1)$]

8. What is the fifth term of the sequence generated by the definition
   $a_n = 10 - 4(n-1)$?

   (1) 18   (3) 2

   (2) 12   (4) -6

9. What definition would produce the sequence 4, 13, 22, 31, ...?

   (1) $a_n = 4 + 9n$ (3) $a_n = 9 + 4n$

   (2) $a_n = 4 + 9(n-1)$ (4) $a_n = 9 + 4(n-1)$

10. Which expression could be used to find the 20th term of the sequence
    5, 9, 13, 17, 21, ...?

    (1) $5 + 4(20)$ (3) $4 + 5(20)$

    (2) $5 + 4(19)$ (4) $4 + 5(19)$

[handwritten:
$$a_n = a_1 + d(n-1)$$
$$a_{154} = a_1 + 7(154-1)$$
$$a_{154} = a_1 + 7 \cdot 153$$
$$a_{154} = a_1 + 1071$$
$$a_{154} = a_1$$
]

Solutions for Practice Exercises: Topic 11

1. Since $7 = 3 + 4$ and $11 = 7 + 4$ and $15 = 11 + 4$, this is an increasing arithmetic sequence.

   The correct choice is (**1**).

2. Since $8 = 2 \cdot 4$ and $16 = 2 \cdot 8$ and $32 = 2 \cdot 16$, this is an increasing geometric sequence.

   The correct choice is (**3**).

3. Each term is equal to $\frac{1}{2}$ multiplied by the previous term. So the next term is $\frac{1}{2} \cdot \frac{5}{2} = \frac{5}{4}$.

   The correct choice is (**1**).

4. $a_8 - a_4 = 33 - 17 = 16$. Since $8 - 4 = 4$, the a_8 term is four terms after the a_4 term. Divide $16 \div 4 = 4$ to get the difference between consecutive terms. So $a_5 = a_4 + 4 = 17 + 4 = 21$.

   The correct choice is (**3**).

5. $a_5 - a_2 = 21 - 36 = -15$. Since $5 - 2 = 3$, the a_5 term is three terms after the a_2 term. Divide $-15 \div 3 = -5$ to get the difference between consecutive terms. So $a_7 = a_5 - 5 - 5 = 21 - 5 - 5 = 11$.

   The correct choice is (**3**).

6. $a_{45} - a_{20} = 272 - 122 = 150$. Since $45 - 20 = 25$, the a_{45} term is 25 terms after the a_{20} term. Divide $150 \div 25 = 6$ to get the difference between consecutive terms. Unlike questions 4 and 5, where the term you are trying to calculate is close to the terms you already have, calculating a_1 is more easily done with the formula from the reference sheet: $a_n = a_1 + d(n - 1)$. Substituting 6 for d and either $n = 45$ or $n = 20$ will work. Here is what the solution looks like for $n = 20$:

$$a_{20} = a_1 + 6(20 - 1)$$
$$122 = a_1 + 6 \cdot 19$$
$$122 = a_1 + 114$$
$$8 = a_1$$

The correct choice is (**1**).

7. $a_{74} - a_{58} = 507 - 395 = 112$. Since $74 - 58 = 16$, the a_{74} term is 16 terms after the a_{58} term. Divide $112 \div 16 = 7$ to get the difference between consecutive terms. Substitute 7 for d and either $n = 74$ or $n = 58$ into the formula $a_n = a_1 + d(n - 1)$ to calculate a_1.

$$a_{58} = a_1 + 7(58 - 1)$$
$$395 = a_1 + 7 \cdot 57$$
$$395 = a_1 + 399$$
$$-4 = a_1$$

Now you can substitute $n = 211$, $d = 7$, and $a_1 = -4$ into the same formula to calculate a_{211}.

$$a_{211} = a_1 + 7(211 - 1)$$
$$a_{211} = -4 + 7 \cdot 210$$
$$a_{211} = -4 + 1,470$$
$$a_{211} = 1,466$$

The correct choice is **(3)**.

8. Substitute 5 for n to get $a_5 = 10 - 4(5 - 1) = 10 - 4(4) = 10 - 16 = -6$.

The correct choice is **(4)**.

9. Substitute 1 for n in each equation. Only the second equation gets $a_1 = 4$. Also, an arithmetic sequence has the form $a_n = a_1 + d(n - 1)$, where d is the common difference and a_1 is the first term. Since $a_1 = 4$ and $d = 9$, the equation is $a_n = 4 + 9(n - 1)$.

The correct choice is **(2)**.

10. An arithmetic sequence with the first term of a_1 and the common difference of d has the equation $a_n = a_1 + d(n - 1)$. a_1 is 5 in this example, and $d = 4$. Therefore, the equation is $a_n = 5 + 4(n - 1)$. Substitute 20 for n to get $20 = 5 + 4(19)$.

The correct choice is **(2)**.

12. Regression Curves

12.1 The Line of Best Fit

A **line of best fit** is a line that comes as close as possible to a set of points on a graph. In the scatterplot below, there is no line that could pass through all 14 points. Of all the possible lines, though, there is one that is a better fit than the others, and this is the line of best fit.

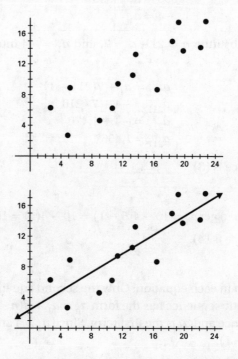

The equation for the line of best fit can be determined quickly on a graphing calculator. The following is a scatterplot and the chart on which it was based.

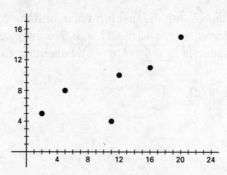

x	y
2	5
5	8
11	4
12	10
16	11
20	15

- Instructions for the TI-84:

Press [STAT] and [1], and enter the *x*-values into L1 and the *y*-values into L2.

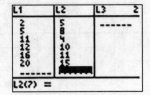

Press [STAT], [right], [4], and [ENTER] for the equation of the line of best fit.

- Instructions for the TI-Nspire:

From the home screen, select the Add Lists & Spreadsheet icon. In the header row, label column A with an "*x*" and column B with a "*y*." Enter the *x*-values into cells A1 to A6. Enter the *y*-values into cells B1 to B6.

Press [menu], [4], [1], and [3] and select *x* in the X List field and *y* in the Y List field

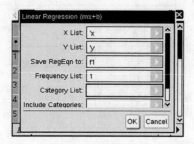

Press the OK button.

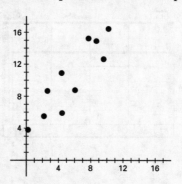

The equation for the line of best fit is $y = 0.477679x + 3.57887$.

12.2 The Correlation Coefficient

The **correlation coefficient**, denoted by the variable r, is a number between -1 and 1. When a line of best fit has a positive slope and passes exactly through each of the points the correlation coefficient is 1. When a line of best fit has a negative slope and passes exactly through each of the points, the correlation coefficient is -1. All correlation coefficients are between -1 and 1.

- The line of best fit for the following scatterplot is close to 1 because it has a positive slope and comes close to the points on the scatterplot. In this case, $r = 0.9$.

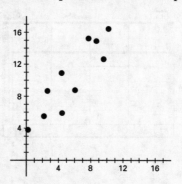

- The line of best fit for the following scatterplot is close to -1 because it has a negative slope and comes close to the points on the scatterplot but not as close as the line in the previous example did to the points on that graph. In this case, the $r = -0.8$.

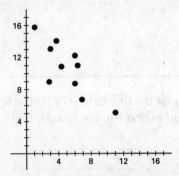

The graphing calculator can display the correlation coefficient. On the TI-84, it will be displayed along with the line of best fit only if the "diagnostics" are turned on. Press [2ND] and [0] to access the catalog, and scroll to the DiagnosticOn command to do this. The TI-Nspire will display the correlation coefficient along with the equation of the line of best fit.

```
CATALOG                    🔼
 DependAuto
 det(
 DiagnosticOff
▶DiagnosticOn
 dim(
 Disp
 DispGraph
```

```
LinReg
 y=ax+b
 a=.4776785714
 b=3.578869048
 r²=.6170415349
 r=.7855199138
```

The correlation coefficient for the example from Section 12.1 was $r = 0.78552$.

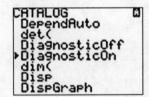

Practice Exercises: Topic 12

1. Calculate the equation for the line of best fit for the following set of data in $y = mx + b$ form. Round m and b to the nearest tenth.

x	y
1	3
2	5
3	4
4	6
5	8

(1) $y = 1.1x + 1.9$

(2) $y = 1.4x + 1.7$

(3) $y = 1.7x + 1.4$

(4) $y = 1.9x + 1.1$

2. Calculate the equation for the line of best fit for the following set of data in $y = mx + b$ form. Round m and b to the nearest tenth.

x	y
10	33
20	20
30	10
40	14
50	6

(1) $y = 34.6x - 0.6$ (3) $y = 31.8x - 0.7$

(2) $y = -0.6x + 34.6$ (4) $y = -0.7x + 31.8$

3. What is the equation for the line of best fit for the points on this scatterplot?

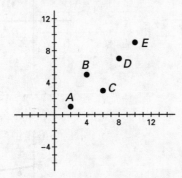

(1) $y = -0.5x + 7$ (3) $y = -0.4x + 0.9$

(2) $y = 0.7x - 0.5$ (4) $y = 0.9x - 0.4$

4. Of these four choices, which line appears to be the best fit for this scatterplot?

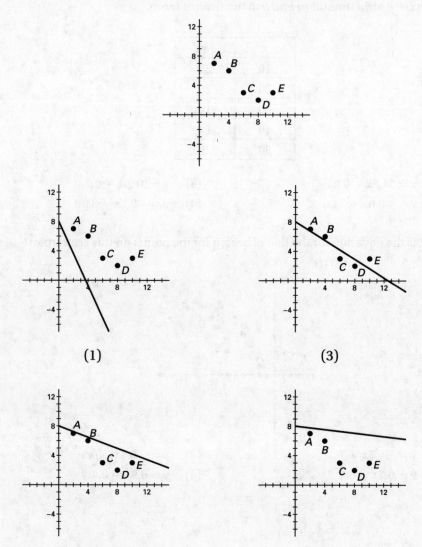

(1)

(2)

(3)

(4)

5. What is the correlation coefficient (r), rounded to the nearest hundredth, for the line of best fit for the data in the table below?

x	y
3	10
6	13
9	27
12	38
15	40

(1) 0.97

(2) 0.96

(3) 0.95

(4) 0.94

6. For which scatterplot is the correlation coefficient closest to 1?

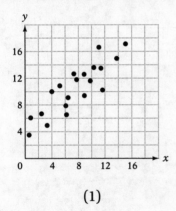

(1)

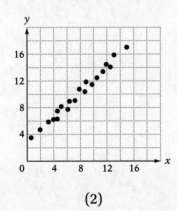

(2)

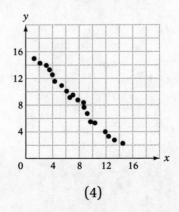

(3)

(4)

7. Of the four choices, which is closest to the correlation coefficient for this scatterplot?

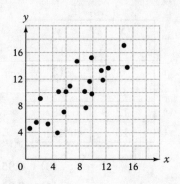

(1)   0.8   (3)   1

(2)   −0.8   (4)   −1

Solutions for Practice Exercises: Topic 12

1. Enter the data into the graphing calculator, and do linear regression to get the equation $y = 1.1x + 1.9$.

   The correct choice is **(1)**.

2. Enter the data into the graphing calculator, and do linear regression to get the equation $y = -0.6x + 34.6$.

   The correct choice is **(2)**.

3. Enter 2, 4, 6, 8, 10 for the x-values and 1, 5, 3, 7, 9 for the y-values into the graphing calculator, and do linear regression. The equation is $y = 0.9x - 0.4$.

   The correct choice is **(4)**.

4. Of the four choices, choice (3) seems to have the line closer to the points than the other three choices.

   The correct choice is **(3)**.

5. Enter the data into the graphing calculator, and do linear regression to get an r-value of 0.97.

   The correct choice is **(1)**.

6. To have an r-value close to $+1$, the points must lie close to a line with a positive slope. Choices (1) and (2) both resemble lines with positive slopes. In choice (2), the points seem to fall closer to a straight line than in choice (1).

   The correct choice is **(2)**.

7. Since this scatterplot resembles a line with a positive slope, choices (2) and (4) can be eliminated because r must be positive. To have an r-value of $+1$, the points would have to lie perfectly on a line, which they do not, so choice (3) can be eliminated too.

   The correct choice is **(1)**.

13. Statistics

13.1 Mean, Median, and Mode

In a set of numbers, the **mean**, also known as the **average**, is the sum of the numbers divided by how many numbers there are. For the set 70, 75, 78, 80, 82, the mean can be calculated as $\dfrac{70 + 75 + 78 + 80 + 82}{5} = \dfrac{385}{5} = 77$.

The **median** is the middle number (or average of the two middle numbers if there are an even amount of numbers) when the numbers are arranged from least to greatest. In the set 70, 75, 78, 80, 82, the median is the number 78. If the set were 70, 75, 78, 80, 82, 85, the median would be the average of 78 and 80, which is $\dfrac{78 + 80}{2} = 79$.

The **mode** is the number that appears most frequently. In the set 70, 75, 78, 80, 80, the mode is 80 since there are two 80s and only one of each of the other numbers.

13.2 First Quartile and Third Quartile

The median is greater than 50% of the numbers in the list. The number that is greater than just 25% of the numbers in the list is called the **first quartile**. The number that is greater than 75% of the numbers in the list is called the **third quartile**. To find the first quartile, find the median of all the numbers less than the median of the list. To find the third quartile, find the median of all the numbers greater than the median of the list. The **interquartile range** is the difference between the third quartile and the first quartile.

For the list 40, 43, 45, 47, 48, 52, 57, 60, 61, 64, 68:

- The numbers less than the median 52 are 40, 43, 45, 47, 48. The median of these five numbers is 45, which is the first quartile.

- The numbers greater than the median 52 are 57, 60, 61, 64, 68. The median of these five numbers is 61, which is the third quartile.

- The interquartile range is $61 - 45 = 16$.

13.3 Box Plots

A box plot is a picture that shows five different metrics: minimum, first quartile, median, third quartile, and maximum. To make a box plot, draw a segment connecting the maximum and the minimum. Then draw a rectangle with sides at the first quartile and third quartile. Finally, draw a vertical line through the rectangle at the median.

- For the following numbers, the box plot looks like this:

    10, 10, 10, 12, 12, 12, 12, 17, 25, 25, 25, 25, 30, 30, 30

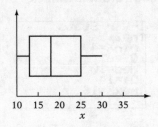

13.4 Using the Graphing Calculator to Determine Maximum, Minimum, Median, First Quartile, and Third Quartile

- For the TI-84:

    The graphing calculator can calculate the five measures of central tendency. First enter all the numbers into L1 by pressing [STAT] and [1] for Edit.

    To find the minimum, first quartile, median, third quartile, and maximum of the seven numbers 10, 4, 8, 12, 6, 16, 14, enter them into L1. Then press [STAT] and [1] for 1-Var Stats and press [ENTER].

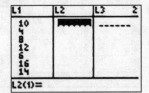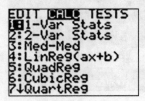

The screen will display

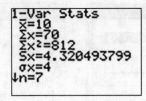

The $\bar{x} = 10$ indicates the mean. The $n = 7$ means that there were seven elements in the list. For the minimum, first quartile, median, third quartile, and maximum, press the down arrow five times.

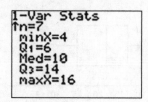

minX is for the minimum, Q1 is for the first quartile, Med is for the median, Q3 is for the third quartile, and maxX is for the maximum element.

■ For the TI-Nspire:

From the home screen, select the Add Lists & Spreadsheet icon. Name column A "x" and fill in cells A1 through A7 with the numbers 10, 4, 8, 12, 6, 16, 14.

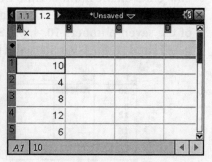

Press [4], [1], and [1] for One-Variable Statistics.

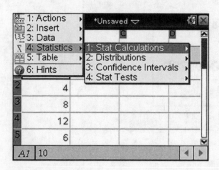
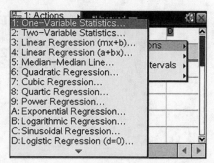

Press [OK] since there is just one list. Set the X1 List to x since that was what the column with the data was named in the spreadsheet.

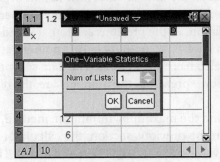 

In cells C2 through C13, the one-variable statistics will be displayed. The median is the $\bar{x}$-bar. n is for the amount of numbers. MinX is the smallest number. Q1X is the first quartile. MedianX is the median. Q3X is the third quartile. MaxX is the largest number.

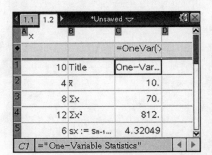

13.5 Standard Deviation

Standard deviation measures how spread out the data are and how far, on average, data are from the mean. When the numbers in a data set are very close together, the data set is said to have a low standard deviation. In the previous section, it was determined that the mean of the seven numbers 10, 4, 8, 12, 6, 16, 14 is 10. Another data set with the numbers 10, 10, 10, 10, 10, 10, 10 also has a mean of 10. The data set with all 10s has a standard deviation of zero. The more spread out the numbers are, the bigger the standard deviation is.

When using the graphing calculator to find the maximum, minimum, median, first quartile, and third quartile, as was done in the previous section, the calculator also displays on that same screen the following values: $S_x = 4.320493799$ and $\sigma_x = 4$. These represent the two different kinds of standard deviation. The S_x is for the **sample standard deviation** (for a small section of a larger group), and the σ_x is for the **population standard deviation** (for the entire larger group). Make sure you know which standard deviation the question is asking for and give that answer!

Practice Exercises: Topic 13

1. Find the mean of this set of numbers: $\{4, 5, 8, 8, 8, 10, 10, 13, 15, 17, 23\}$.

   (1) 8   (3) 10
   (2) 9   (4) 11

2. Find the median of this set of numbers: $\{4, 5, 8, 8, 8, 10, 10, 13, 15, 17, 23\}$.

   (1) 8   (3) 11
   (2) 10   (4) 15

3. Find the interquartile range of this set of numbers: $\{4, 5, 8, 8, 8, 10, 10, 13, 15, 17, 23\}$.

   (1) 7   (3) 9
   (2) 8   (4) 10

4. For the first four days of a five-day vacation, the mean temperature was 80 degrees. What must the temperature be on the fifth day in order for the mean temperature to be 82 degrees?

   (1) 88   (3) 90
   (2) 89   (4) 91

5. For which data set is the median greater than the mean?

   (1) $\{4, 7, 10, 13, 16\}$ (3) $\{8, 9, 10, 11, 12\}$
   (2) $\{8, 9, 10, 18, 19\}$ (4) $\{1, 2, 10, 11, 12\}$

6. What is the mode of the data in this histogram?

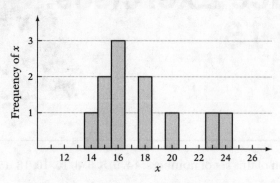

   (1)   23   (3)   15

   (2)   24   (4)   16

7. What is the median of the data in this box plot?

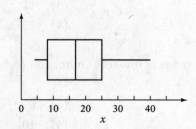

   (1)   17   (3)   8

   (2)   15   (4)   25

8. What is true about this data set: 1, 2, 10, 11, 11?

   (1)   The median is greater than the mean.

   (2)   The median is equal to the mean.

   (3)   The median is less than the mean.

   (4)   The median is greater than the mode.

9. For which data set is the interquartile range equal to 0?

    (1)   2, 3, 6, 6, 6, 7, 8

    (2)   2, 6, 6, 6, 6, 6, 8

    (3)   1, 2, 3, 6, 7, 8, 9

    (4)   2, 6, 6, 6, 6, 7, 8

10. What is the sample standard deviation, rounded to the nearest hundredth, of this data set: 20, 25, 28, 30, 32, 40?

    (1)   29.17

    (2)   6.77

    (3)   6.18

    (4)   20

Solutions for Practice Exercises: Topic 13

1. To get the mean, add the 11 values and divide by 11: $(4 + 5 + 8 + 8 + 8 + 10 + 10 + 13 + 15 + 17 + 23)/11 = 121/11 = 11$.

   The correct choice is **(4)**.

2. The median is the middle number after the numbers have been arranged from least to greatest. These numbers are already arranged from least to greatest, so the middle number is the sixth number, which is 10.

   The correct choice is **(2)**.

3. The interquartile range is the difference between the first quartile and the third quartile. The median is the sixth number, so the first quartile is the median of the first five numbers. The first five numbers are 4, 5, 8, 8, 8, with a median of 8. The third quartile is the median of the last five numbers. The last five numbers are 10, 13, 15, 17, 23, with a median of 15. The first quartile is 8. The third quartile is 15. The interquartile range is $15 - 8 = 7$.

   The correct choice is **(1)**.

4. The sum of a set of numbers is the product of the average and the amount of numbers. If the first four days had an average of 80, the sum of all four temperatures was 320. If the first five days need to have an average of 82, the sum of all five temperatures needs to be $82 \cdot 5 = 410$. The fifth temperature needs to be $410 - 320 = 90$.

   The correct choice is **(3)**.

5. The median for all four data sets is 10. The mean of the numbers in choice (4) is $(1 + 2 + 10 + 11 + 12)/5 = 7.2$, which is less than the median. All the other choices have the mean greater than or equal to the median.

   The correct choice is **(4)**.

6. The mode is the most frequent number. In a histogram, the most frequent number is the tallest bar, which is 16 in this example.

   The correct choice is **(4)**.

7. The vertical line in the middle of the rectangle represents the median in the box plot. Since this is at 17, the median is 17.

   The correct choice is **(1)**.

8. The median of the data set is 10. The mode is 11. The mean is 7. The median is greater than the mean.

   The correct choice is **(1)**.

9. The interquartile range is 0 if the first quartile is equal to the third quartile. This is the case for choice (2), where both the first quartile and the third quartile are 6.

   The correct choice is **(2)**.

10. Enter the numbers into the graphing calculator, and find the one-variable statistics menu. It should output that $S_x = 6.77$ and $\sigma_x = 6.18$. Since the question asks about the *sample* standard deviation, use $S_x = 6.77$.

    The correct choice is **(2)**.

Glossary of Terms

Addition property of equality A property of algebra that states that when equal values are added to both sides of a true equation, the equation continues to be true. To solve the equation $x - 2 = 5$, add 2 to both sides of the equation by using the addition property of equality.

Arithmetic sequence A number sequence in which the difference between two consecutive terms is a constant. The sequence 2, 5, 8, 11, 14, . . . is an arithmetic sequence because the difference between consecutive terms is always 3.

Average rate of change A measure of how much the output of a function changes for each unit of change in the input. The formula for the average rate of change over the interval a to b is $\dfrac{f(b) - f(a)}{b - a}$.

Axis of symmetry An imaginary vertical line that passes through the vertex of a parabola. The equation for the axis of symmetry of a parabola defined by $y = ax^2 + bx + c$ is $x = \dfrac{-b}{2a}$.

Base The number being raised to a power in an exponential expression. In the expression $2 \cdot 3^x$, the 3 is the base.

Binomial A polynomial with only two terms. $3x + 5$ is a binomial.

Box plot A graphical way to summarize data. The five numbers represented by the minimum, first quartile, median, third quartile, and maximum are graphed on a number line. A line segment connects the minimum to the first quartile. A rectangle is drawn around the first quartile and third quartile with a vertical line at the median. A line segment connects the third quartile to the maximum.

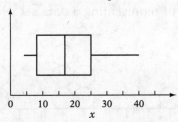

Closed form defined sequence A formula that defines the nth term of a sequence. The formula $a_n = 3 + 2(n - 1)$ is a closed form definition of a sequence. To get the 50th term of the sequence, substitute 50 for n in the definition.

Coefficient A number multiplied by a variable expression. In the expression $5x + 2$, 5 is the coefficient of x.

Common difference In an arithmetic sequence, the difference between consecutive terms. The common difference in the sequence 2, 5, 8, 11, 14, . . . is 3.

Common ratio In a geometric sequence, the ratio between consecutive terms. The common ratio in the sequence 2, 6, 18, 54, 162, . . . is 3 since $162/54 = 54/18 = 18/6 = 6/2 = 3$.

Commutative property of addition or multiplication The law from arithmetic that states that the order in which two numbers are added or multiplied does not matter. Because of the commutative property of addition, $5 + 2 = 2 + 5$.

Completing the square A method of solving a quadratic equation that involves turning one side of the equation into a perfect square trinomial.

Constant A number that does not have a variable part. In the expression $5x + 2$, 2 is a constant.

Correlation coefficient A number represented by r that measures how well a curve of best fit matches the points in a scatterplot. When the correlation coefficient is very close to 1 or to -1, the curve is a very good fit.

Degree of a polynomial The exponent on the highest power of a polynomial. In the polynomial $x^3 - 2x^2 + 5x - 2$, the degree is 3 since the highest power is a 3.

Difference of perfect squares Factoring a quadratic binomial in the form $x^2 - a^2$ into $(x - a)(x + a)$. For example, $x^2 - 9 = (x - 3)(x + 3)$.

Distributive property of multiplication over addition The rule that allows expressions of the form $a(b + c)$ to become $a \cdot b + a \cdot c$. For example, $2(3x + 5) = 6x + 10$.

Division property of equality A property of algebra that states that when both sides of a true equation are divided by the same non-zero number, the equation continues to be true. To solve the equation $2x = 8$, divide both sides of the equation by 2 using the division property of equality.

Domain The numbers that can be input into a function. When the function is defined as a set of ordered pairs or as a graph, the domain is the set of x-coordinates.

Dot plot A graphical way of representing a data set where each piece of data is represented with a dot.

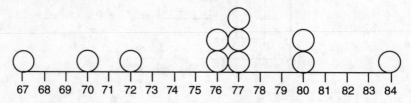

Elimination method A way of solving a system of linear equations by combining the two equations in such a way as to eliminate one of the variables. In the set of equations,

$$x + 2y = 12$$
$$3x - 2y = -4$$

the y-variable is eliminated by adding the two equations together.

Equation Two mathematical expressions with an equal sign between them. $3x + 2 = 8$ is an equation.

Exponential equation An equation in which the variable is an exponent. $2 \cdot 3^x = 18$ is an exponential equation.

Exponential function A function in which the variable is an exponent. $f(x) = 2 \cdot 3^x$ is an exponential function.

Expression Numbers and variables that are combined with the operations from math.

Factoring a polynomial Finding two polynomials that can be multiplied to become another polynomial. The polynomial $x^2 + 5x + 6$ can be factored into $(x + 2)(x + 3)$.

Factors The polynomials that evenly divide into a polynomial. The factors of $x^2 + 5x + 6$ are $(x + 2)$ and $(x + 3)$.

First quartile The number in a data set that is bigger than 25% of the numbers in the set.

FOIL A way of multiplying two binomials of the form $(a + b)(c + d)$ where (F)irst the a and c are multiplied, then the (O)uters a and d are multiplied, then the (I)nners b and c are multiplied, and finally the (L)asts b and d are multiplied. Then the four results are added together. The product of $(x + 2)$ and $(x + 3)$ is $x^2 + 3x + 2x + 6 = x^2 + 5x + 6$ by this method.

Function Something that takes numbers as inputs and outputs numbers. Functions are often labeled with the letters f or g. The notation $f(2) = 7$ means that when the number 2 is input into function f, it outputs the number 7.

Geometric sequence A number sequence in which the ratio between two consecutive terms (what you get when you divide one term by the term before it) is a constant. The sequence 2, 6, 18, 54, 162, . . . is a geometric sequence because the ratio between consecutive terms is always 3.

Graph A visual way to describe the solution set to an equation. Each solution to the equation corresponds to an ordered pair that is graphed as a point on the coordinate plane. Each of the ordered pairs that satisfies an equation produces a point on the coordinate plane, and the collection of all the points is the graph of the equation.

Greatest common factor The largest expression that divides evenly into two or more monomials. The greatest common factor of $6x^2$ and $8x^3$ is $2x^2$.

Growth rate In the exponential expression $a \cdot b^x$, the b is the growth rate. For example, in the equation $y = 500 \cdot 1.05^x$, the growth rate is 1.05.

Histogram A way of representing data with repeated values. Each value is represented by a bar whose height corresponds to the number of times that value is repeated. There are no spaces between the bars.

Increasing A function is increasing on an interval if making the input value larger also makes the y output larger. On a graph, an increasing function "goes up" from left to right.

Inequality Like an equation, but there is a $<$, $>$, $\leq$, or $\geq$ sign between the two expressions. $x + 2 > 5$ is a one-variable inequality. $y \leq 2x + 6$ is a two-variable inequality.

Interquartile range The difference between the number that is the third quartile of a data set and the number that is the first quartile of a data set.

Isolating a variable A variable is isolated when it is by itself on one side of an equation. In the equation $x + 2 = 5$, the x is not yet isolated. Subtracting 2 from both sides of the equation transforms the original equation into $x = 3$ with the x now isolated.

Like terms Terms that have the same variable part. They can be combined by adding or subtracting. $2x^2$ and $3x^2$ are like terms. $2x^2$ and $3x^3$ are not like terms.

Line of best fit A line that comes closest to the set of points in a scatterplot.

Linear equation An equation in which the greatest exponent is 1. The equation $2x + 3 = 7$ is a linear equation.

Linear function A function in which the greatest exponent is 1. The function $f(x) = 2x + 3$ is a linear function.

Mean The average of the numbers in a data set. Calculate the mean by adding all the numbers and dividing the total by the amount of numbers in the set.

Median The middle number in a data set after it has been arranged from least to greatest. If there are an even amount of numbers in the data set, the median is found by adding the two middle numbers and dividing by 2.

Mode The most frequent number in a data set.

Monomial A mathematical expression that has a coefficient and/or a variable part. $3x^2$ is a monomial. $3 + x^2$ is not a monomial.

Multiplication property of equality A property of algebra that states that when equal values are multiplied by both sides of a true equation, the equation continues to be true. To solve the equation $(1/2)x = 5$, multiply both sides of the equation by 2 using the multiplication property of equality.

Ordered pair Two numbers written in the form (x, y). An ordered pair can be a solution to a two-variable equation. For example, $(2, 5)$ is one solution to the equation $y = 2x + 1$. Ordered pairs can be graphed on the coordinate axes by locating the point with the x-coordinate equal to the x-value and the y-coordinate equal to the y-value.

Parabola A U-shaped curve that is the graph of the solution set of a quadratic equation.

Perfect square trinomial A quadratic trinomial of the form $x^2 + bx + (b/2)^2$, which can be factored into $(x + b/2)^2$. For example, $x^2 + 6x + 9 = (x + 3)^2$.

Piecewise function A function that has multiple rules for determining output values from input values, depending on what the input values are. If the function $f(x)$ is defined as

$$f(x) = \begin{array}{ll} 2x + 1 & \text{if } x < 0 \\ x^2 & \text{if } x \geq 0 \end{array}$$

then $f(-3) = 2(-3) + 1 = -5$ and $f(5) = 5^2 = 25$.

Polynomial The sum of one or more monomials. Each term of the polynomial has the form ax^n. $3x^4 + 2x^3 - 3x^2 + 6x - 1$ is a polynomial.

Quadratic equation An equation in which the highest power on a variable is a 2. $x^2 + 5x + 6 = 0$ is a quadratic equation.

Quadratic formula The formula $x = (-b \pm \sqrt{(b^2 - 4ac)})/2a$. This formula can be used to find the two solutions to the quadratic equation $ax^2 + bx + c = 0$.

Quadratic function A function in which the highest power on a variable is 2. $f(x) = x^2 + 5x + 6$ is a quadratic function.

Quadratic polynomial A polynomial in which the highest power on a variable is 2. $x^2 + 5x + 6$ is a quadratic polynomial.

Range The set of values that can be output from a function is the range of that function.

Rationalizing the denominator A process by which a fraction that has an irrational denominator, like $\frac{5}{\sqrt{3}}$, is converted to an equivalent fraction that does not have an irrational denominator.

Regression Finding a curve that best fits a scatterplot. Three types of regression are linear, quadratic, and exponential.

Residual plot A set of points that represents how far points on a graph deviate from a curve of best fit.

Roots The roots of an equation are the values that solve that equation. The roots of $x^2 + 5x + 6 = 0$ are -3 and -2.

Sequence A list of numbers that usually has some kind of pattern.

Slope The slope of a line is a measure of the line's steepness. The equation for the slope of a line that passes through the two points (x_1, y_1) and (x_2, y_2) is $m = \dfrac{y_2 - y_1}{x_2 - x_1}$.

Slope-intercept form An equation in the form $y = mx + b$ where m and b are numbers in slope-intercept form. $y = 2x - 1$ is in slope-intercept form. When a two-variable equation is in slope-intercept form, the graph of the equation has a y-intercept of $(0, b)$ and a slope of m.

Solution set The set of numbers or ordered pairs that satisfies an equation. The solution set of $x + 2 = 5$ is $\{3\}$. The solution set of $x + y = 10$ has an infinite number of ordered pairs in its solution set, including $(2, 8)$, $(3, 7)$, and $(4, 6)$.

Standard deviation A measure of how spread out the numbers in a data set are. If all the numbers are the same, the standard deviation is zero. Otherwise, the standard deviation is greater than zero. The more spread out the numbers are, the higher the standard deviation is. There are two kinds of standard deviation: sample standard deviation (S_x) and population standard deviation (σ_x).

Substitution method A method for solving a system of equations in which one variable is isolated in one of the equations and the expression equal to that variable is substituted for it in the other equation.

Subtraction property of equality A property of algebra that states that when equal values are subtracted from both sides of a true equation, the equation continues to be true. To solve the equation $x + 2 = 5$, subtract 2 from both sides of the equation by using the subtraction property of equality.

System of equations Two or more equations with two or more unknowns to solve for. An example of a system of equations with a solution of $(8, 2)$ is

$$x + y = 10$$
$$x - y = 6$$

Third quartile The number in a data set that is greater than 75% of the numbers in the set.

Translation A translation of a graph is when the points on it are each shifted the same amount in the same direction. Examples of translations are vertical translations, horizontal translations, and combinations of vertical and horizontal translations.

Trinomial A polynomial with three terms. The polynomial $x^2 + 5x + 6$ is a trinomial.

Variable A letter, often an x, y, or z, that represents a value in a mathematical expression. In an algebraic equation, the variable is often the unknown that needs to be solved for.

Vertex The turning point of a parabola is its vertex. If the parabola opens upward, the vertex is the minimum point. If the parabola opens downward, the vertex is the maximum point.

Vertical line test A way of testing to see if a graph can represent a function. If at least one vertical line can pass through at least two points on the graph, the graph fails the vertical line test and cannot represent a function. If there are no vertical lines that can pass through at least two points, then the graph can be the graph of a function.

***x*-intercept** The location where a curve crosses the x-axis. The y-coordinate of the x-intercept is 0.

***y*-intercept** The location where a curve crosses the y-axis. The x-coordinate of the y-intercept is 0. In slope-intercept form, $y = mx + b$, the y-intercept is located at $(0, b)$.

Zeros The zeros of a function f are the numbers that can be input into the function so that 0 is output from the function. For example, the function $f(x) = 2x - 6$ has the number 3 as its only zero since $f(3) = 2(3) - 6 = 6 - 6 = 0$.

Regents Exams, Answers, and Self-Analysis Charts

June 2019 Exam
Algebra I

High School Math Reference Sheet

Conversions

1 inch = 2.54 centimeters

1 meter = 39.37 inches

1 mile = 5280 feet

1 mile = 1760 yards

1 mile = 1.609 kilometers

1 kilometer = 0.62 mile

1 pound = 16 ounces

1 pound = 0.454 kilogram

1 kilogram = 2.2 pounds

1 ton = 2000 pounds

1 cup = 8 fluid ounces

1 pint = 2 cups

1 quart = 2 pints

1 gallon = 4 quarts

1 gallon = 3.785 liters

1 liter = 0.264 gallon

1 liter = 1000 cubic centimeters

Formulas

Triangle	$A = \frac{1}{2}bh$
Parallelogram	$A = bh$
Circle	$A = \pi r^2$
Circle	$C = \pi d$ or $C = 2\pi r$

Formulas (continued)

General Prisms	$V = Bh$
Cylinder	$V = \pi r^2 h$
Sphere	$V = \frac{4}{3}\pi r^3$
Cone	$V = \frac{1}{3}\pi r^2 h$
Pyramid	$V = \frac{1}{3}Bh$
Pythagorean Theorem	$a^2 + b^2 = c^2$
Quadratic Formula	$x = \dfrac{-b \pm \sqrt{b^2 - 4ac}}{2a}$
Arithmetic Sequence	$a_n = a_1 + (n - 1)d$
Geometric Sequence	$a_n = a_1 r^{n-1}$
Geometric Series	$S_n = \dfrac{a_1 - a_1 r^n}{1 - r}$ where $r \neq 1$
Radians	1 radian $= \frac{180}{\pi}$ degrees
Degrees	1 degree $= \frac{\pi}{180}$ radians
Exponential Growth/Decay	$A = A_0 e^{k(t - t_0)} + B_0$

PART I

Answer all 24 questions in this part. Each correct answer will receive 2 credits. No partial credit will be allowed. For each statement or question, write in the space provided the numeral preceding the word or expression that best completes the statement or answers the question.   [48 credits]

1. The expression $w^4 - 36$ is equivalent to

(1) $(w^2 - 18)(w^2 - 18)$ (3) $(w^2 - 6)(w^2 - 6)$

(2) $(w^2 + 18)(w^2 - 18)$ (4) $(w^2 + 6)(w^2 - 6)$ 1

2. If $f(x) = 4x + 5$, what is the value of $f(-3)$?

(1) -2 (3)   17

(2) -7 (4)   4   2 _2_

3. Which relation is *not* a function?

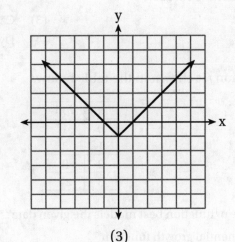

x	y
-10	-2
-6	2
-2	6
1	9
5	13

       (1)   (3)

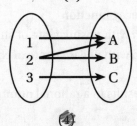

$3x + 2y = 4$

       (2)   (4)   3 _2_

4. Given: $f(x) = (x-2)^2 + 4$

$\qquad g(x) = (x-5)^2 + 4$

When compared to the graph of $f(x)$, the graph of $g(x)$ is

(1) shifted 3 units to the left

(2) shifted 3 units to the right

(3) shifted 5 units to the left

(4) shifted 5 units to the right

4 ___2___

5. Students were asked to write $6x^5 + 8x - 3x^3 + 7x^7$ in standard form.

Shown below are four student responses.

Anne: $7x^7 + 6x^5 - 3x^3 + 8x$

Bob: $-3x^3 + 6x^5 + 7x^7 + 8x$

Carrie: $8x + 7x^7 + 6x^5 - 3x^3$

Dylan: $8x - 3x^3 + 6x^5 + 7x^7$

Which student is correct?

(1) Anne   (3) Carrie

(2) Bob   (4) Dylan

5 ___2___

6. The function f is shown in the table below.

x	f(x)
0	1
1	3
2	9
3	27

Which type of function best models the given data?

(1) exponential growth function

(2) exponential decay function

(3) linear function with positive rate of change

(4) linear function with negative rate of change

6 ___2___

7. Which expression results in a rational number?

(1) $\sqrt{2} \cdot \sqrt{18}$ ✗   (3) $\sqrt{2} + \sqrt{2}$

(2) $5 \cdot \sqrt{5}$ (4) $3\sqrt{2} + 2\sqrt{3}$

7 ___2___

8. A polynomial function is graphed below.

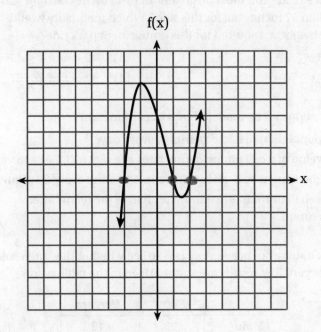

f(x)

Which function could represent this graph?

(1) $f(x) = (x + 1)(x^2 + 2)$

(2) $f(x) = (x - 1)(x^2 - 2)$

(3) $f(x) = (x - 1)(x^2 - 4)$

(4) $f(x) = (x + 1)(x^2 + 4)$

8 _2_

9. When solving $p^2 + 5 = 8p - 7$, Kate wrote $p^2 + 12 = 8p$. The property she used is

(1) the associative property

(2) the commutative property

(3) the distributive property

(4) the addition property of equality

9 _2_

10. David wanted to go on an amusement park ride. A sign posted at the entrance read "You must be greater than 42 inches tall and no more than 57 inches tall for this ride." Which inequality would model the height, x, required for this amusement park ride?

   (1) $42 < x \leq 57$ (3) $42 < x$ or $x \leq 57$

   (2) $42 > x \geq 57$ (4) $42 > x$ or $x \geq 57$ 10 _____

11. Which situation can be modeled by a linear function?

   (1)   The population of bacteria triples every day.

   (2)   The value of a cell phone depreciates at a rate of 3.5% each year.

   (3)   An amusement park allows 50 people to enter every 30 minutes.

   (4)   A baseball tournament eliminates half of the teams after each round.   11 _____

12. Jenna took a survey of her senior class to see whether they preferred pizza or burgers. The results are summarized in the table below.

	Pizza	Burgers
Male	23	42
Female	31	26

Of the people who preferred burgers, approximately what percentage were female?

   (1)   21.3   (3)   45.6

   (2)   38.2   (4)   61.9   12 _____

13. When $3a + 7b > 2a - 8b$ is solved for a, the result is

   (1) $a > -b$ (3) $a < -15b$

   (2) $a < -b$ (4) $a > -15b$ 13 _____

$3a > 2a - 15b$

$a > -15b$

14. Three functions are shown below.

A: $g(x) = -\frac{3}{2}x + 4$

B: $f(x) = (x+2)(x+6)$

C:

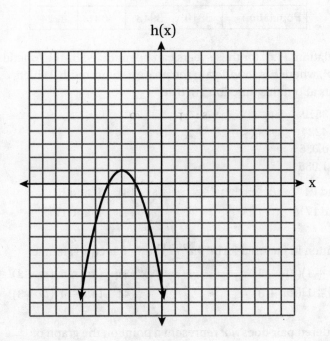

Which statement is true?

(1) B and C have the same zeros. ✗

(2) A and B have the same y-intercept. ✗

(3) B has a minimum and C has a maximum.

(4) C has a maximum and A has a minimum.

14 _2_

15. Nicci's sister is 7 years less than twice Nicci's age, a. The sum of Nicci's age and her sister's age is 41. Which equation represents this relationship?

(1) $a + (7 - 2a) = 41$

(2) $a + (2a - 7) = 41$

(3) $2a - 7 = 41$

(4) $a = 2a - 7$

15 _2_

16. The population of a small town over four years is recorded in the chart below, where 2013 is represented by $x = 0$. [Population is rounded to the nearest person.]

Year	2013	2014	2015	2016
Population	3810	3943	4081	4224

The population, $P(x)$, for these years can be modeled by the function $P(x) = ab^x$, where b is rounded to the nearest thousandth. Which statements about this function are true?

I.　$a = 3810$
II.　$a = 4224$
III.　$b = 0.035$
IV.　$b = 1.035$

(1)　I and III　　　　　　　　(3)　II and III
(2)　I and IV　　　　　　　　(4)　II and IV　　　　16 __2__

17. When written in factored form, $4w^2 - 11w - 3$ is equivalent to

(1)　$(2w + 1)(2w - 3)$　　　　(3)　$(4w + 1)(w - 3)$
(2)　$(2w - 1)(2w + 3)$　　　　(4)　$(4w - 1)(w + 3)$　　17 _____

18. Which ordered pair does *not* represent a point on the graph of $y = 3x^2 - x + 7$?

(1)　$(-1.5, 15.25)$　　　　　　(3)　$(1.25, 10.25)$
(2)　$(0.5, 7.25)$　　　　　　　(4)　$(2.5, 23.25)$　　　18 _____

19. Given the following three sequences:

I.　$2, 4, 6, 8, 10, \ldots$
II.　$2, 4, 8, 16, 32, \ldots$
III.　$a, a + 2, a + 4, a + 6, a + 8, \ldots$

Which ones are arithmetic sequences?

(1)　I and II, only　　　　　　(3)　II and III, only
(2)　I and III, only　　　　　　(4)　I, II, and III　　　19 __3__

20. A grocery store sells packages of beef. The function $C(w)$ represents the cost, in dollars, of a package of beef weighing w pounds. The most appropriate domain for this function would be

(1) integers

(2) rational numbers

(3) positive integers

(4) positive rational numbers

20 _____ 2

21. The roots of $x^2 - 5x - 4 = 0$ are

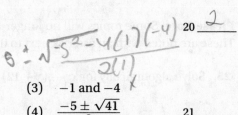

(1) 1 and 4

(3) −1 and −4

(2) $\dfrac{5 \pm \sqrt{41}}{2}$

(4) $\dfrac{-5 \pm \sqrt{41}}{2}$

21 _____

22. The following table shows the heights, in inches, of the players on the opening-night roster of the 2015−2016 New York Knicks.

84	80	87	75	77	79	80	74	76	80	80	82	82

The population standard deviation of these data is approximately

(1) 3.5

(3) 79.7

(2) 13

(4) 80

22 _____

23. A population of bacteria can be modeled by the function $f(t) = 1000(0.98)^t$, where t represents the time since the population started decaying, and $f(t)$ represents the population of the remaining bacteria at time t. What is the rate of decay for this population?

(1) 98%

(3) 0.98%

(2) 2%

(4) 0.02%

23 _____

24. Bamboo plants can grow 91 centimeters per day. What is the approximate growth of the plant, in inches per hour?

(1) 1.49

(3) 9.63

(2) 3.79

(4) 35.83

24 _____

PART II

Answer all 8 questions in this part. Each correct answer will receive 2 credits. Clearly indicate the necessary steps, including appropriate formula substitutions, diagrams, graphs, charts, etc. For all questions in this part, a correct numerical answer with no work shown will receive only 1 credit. [16 credits]

Please Note: Some topics will no longer be tested on future Regents Algebra I exams. These are noted with an asterisk next to the specific questions.

25. Solve algebraically for x: $-\frac{2}{3}(x+12) + \frac{2}{3}x = -\frac{5}{4}x + 2$

$$-\frac{2}{3}x - 8 + \frac{2}{3}x = -\frac{5}{4}x + 2$$

$$-8 = -\frac{5}{4}x + 2$$

$$\frac{5}{4}x = 10$$

$$x = \frac{1}{2}$$

26. If $C = G - 3F$, find the trinomial that represents C when
$F = 2x^2 + 6x - 5$ and $G = 3x^2 + 4$.

$C = (3x^2 + 4) - 3(2x^2 + 6x - 5)$

$C = 3x^2 + 4 - 6x^2 - 18x + 15$

$C = -3x^2 - 18x + 19$

2

27. Graph the following piecewise function on the set of axes below.

$$f(x) = \begin{cases} |x|, & -5 \le x < 2 \\ -2x + 10, & 2 \le x \le 6 \end{cases}$$

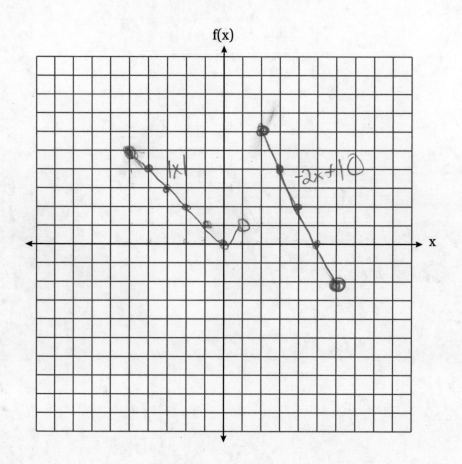

28. Solve $5x^2 = 180$ algebraically.

$$x^2 = \frac{180}{5}$$

$$x^2 = \sqrt{36}$$

$$x = 6, ^-6$$

29. A blizzard occurred on the East Coast during January, 2016. Snowfall totals from the storm were recorded for Washington, D.C. and are shown in the table below.

Washington, D.C.	
Time	**Snow** (inches)
1 a.m.	1
3 a.m.	5
6 a.m.	11
12 noon	33
3 p.m.	36

Which interval, 1 a.m. to 12 noon or 6 a.m. to 3 p.m., has the greatest rate of snowfall, in inches per hour? Justify your answer.

$$1am - 12: \frac{33-1}{12-1} = \frac{32}{11} \approx 2.91$$

$$6am - 3pm: \frac{36-11}{6-3} = \frac{25}{3} \approx 8.\overline{3}$$

6am - 3pm has the greatest rate of snowfall

30. The formula for the volume of a cone is $V = \frac{1}{3}\pi r^2 h$. Solve the equation for h in terms of V, r, and π.

$$v \cdot \frac{1}{3}\pi \div r^2 = h$$

31 Given the recursive formula:

$$a_1 = 3$$
$$a_n = 2(a_{n-1} + 1)$$

State the values of a_2, a_3, and a_4 for the given recursive formula.

$$a_2 = 2(a_{2-1} + 1)$$

$$a_2 = 2(a_1 + 1)$$

$$a_2 = 2(3+1)$$

$$a_2 = 2 \cdot 4$$

$$a_2 = 8$$

$$a_3 = 18$$

$$a_4 = 38$$

32. Determine and state the vertex of $f(x) = x^2 - 2x - 8$ using the method of completing the square.

$$x^2 - 2x = f(x) + 8$$

$$x^2 - 2x + 1 = f(x) + 9$$

$$(x-1)^2 = f(x) + 9$$

$$(x-1)^2 - 9 = f(x)$$

$$(1, -9)$$

PART III

Answer all 4 questions in this part. Each correct answer will receive 4 credits. Clearly indicate the necessary steps, including appropriate formula substitutions, diagrams, graphs, charts, etc. For all questions in this part, a correct numerical answer with no work shown will receive only 1 credit.   [16 credits]

33. A school plans to have a fundraiser before basketball games selling shirts with their school logo. The school contacted two companies to find out how much it would cost to have the shirts made. Company *A* charges a $50 set-up fee and $5 per shirt. Company *B* charges a $25 set-up fee and $6 per shirt.

Write an equation for Company *A* that could be used to determine the total cost, *A*, when *x* shirts are ordered. Write a second equation for Company *B* that could be used to determine the total cost, *B*, when *x* shirts are ordered.

$$A = 50 + 5x$$

$$B = 25 + 6x$$

Determine algebraically and state the *minimum* number of shirts that must be ordered for it to be cheaper to use Company *A*.

Plug into calculator
look at the table

you would have to buy
at least 26 shirts for company
a to be cheaper

34. Graph $y = f(x)$ and $y = g(x)$ on the set of axes below.

$$f(x) = 2x^2 - 8x + 3$$
$$g(x) = -2x + 3$$

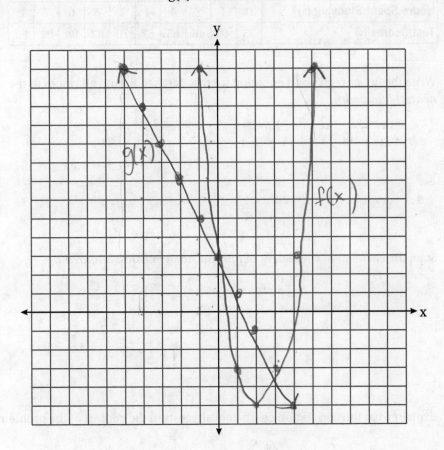

Determine and state all values of x for which $f(x) = g(x)$.

35. The table below shows the number of hours ten students spent studying for a te
and their scores.

Hours Spent Studying (x)	0	1	2	4	4	4	6	6	7	8
Test Scores (y)	35	40	46	65	67	70	82	88	82	95

Write the linear regression equation for this data set. Round all values to the
nearest hundredth.

$$y = 7.79x + 34.27$$

State the correlation coefficient of this line, to the *nearest hundredth.*

.98

Explain what the correlation coefficient suggests in the context of the problem.

strong positive
correlation

36. A system of inequalities is graphed on the set of axes below.

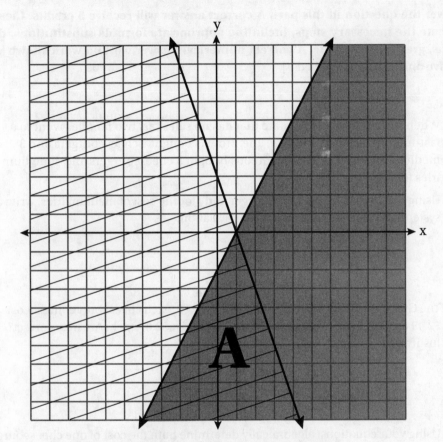

State the system of inequalities represented by the graph.

$$y \le 2x - 2$$
$$y \le 3x + 3$$

State what region *A* represents.

the solution
set

State what the entire gray region represents.

$$y \le 2x - 2$$

PART IV

Answer the question in this part. A correct answer will receive 6 credits. Clearly indicate the necessary steps, including appropriate formula substitutions, diagrams, graphs, charts, etc. A correct numerical answer with no work shown will receive only 1 credit.   [6 credits]

37. When visiting friends in a state that has no sales tax, two families went to a fast-food restaurant for lunch. The Browns bought 4 cheeseburgers and 3 medium fries for $16.53. The Greens bought 5 cheeseburgers and 4 medium fries for $21.11.

    Using c for the cost of a cheeseburger and f for the cost of medium fries, write a system of equations that models this situation.

    $$B: 4c + 3f = 16.53$$
    $$G. 5c + 4f = 21.11$$

    The Greens said that since their bill was $21.11, each cheeseburger must cost $2.49 and each order of medium fries must cost $2.87 each. Are they correct? Justify your answer.

    $$-16c - 12f = -66.12 \qquad NO$$
    $$15c + 12f = 63 + 33$$
    $$-c = -2.79 \qquad c = 2.79$$

    Using your equations, algebraically determine both the cost of one cheeseburger and the cost of one order of medium fries.

    $$4(2.79) + 3f = 16.53$$

    $$11.16 + 3f = 16.53$$
    $$3f = 5.37$$
    $$f = 1.79$$

    $$c = 2.79$$

    6

Answers June 2019

Algebra I

Answer Key

PART I

1. (4)	5. (1)	9. (4)	13. (4)	17. (3)	21. (2)
2. (2)	6. (1)	10. (1)	14. (3)	18. (3)	22. (1)
3. (4)	7. (1)	11. (3)	15. (2)	19. (2)	23. (2)
4. (2)	8. (3)	12. (2)	16. (2)	20. (4)	24. (1)

PART II

25. $x = 8$

26. $-3x^2 - 18x + 19$

27.

28. $\{-6, 6\}$

29. 1 a.m. to 12 noon because
$$2\frac{10}{11} > 2\frac{7}{9}$$

30. $h = \dfrac{3V}{\pi r^2}$

31. 8, 18, 38

32. $(1, -9)$

PART III

33. $A = 5x + 50$, $B = 6x + 25$; $x = 26$

34.

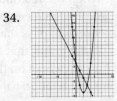

$x = 0, x = 3$

35. $y = 7.79x + 34.27$, $r = 0.98$

36. $y \le 2x - 2$, $y < -3x + 3$; A represents the solution set of the system of inequalities. The gray region is the solution set to $y \le 2x - 2$.

PART IV

37. $4c + 3f = 16.53$, $5c + 4f = 21.11$. They are not correct because it would be $23.93, not $21.11; one cheeseburger is $2.79, and one order of fries is $1.79.

In **Parts II–IV**, you are required to show how you arrived at your answers. For sample methods of solutions, see the *Answer Explanations* section.

Answer Explanations

Part I

1. $w^4 - 36$ can be rewritten as $(w^2)^2 - 6^2$. This can be factored with the factoring pattern $a^2 - b^2 = (a - b)(a + b)$.

$$w^4 - 36 = (w^2)^2 - 6^2 = (w^2 - 6)(w^2 + 6)$$

The correct choice is **(4)**.

2. $f(-3) = 4(-3) + 5 = -12 + 5 = -7$

The correct choice is **(2)**.

3. Testing choice (1): When a relation is described in a table, it is only a function if there are no repeats in the x column. Since there are no repeats in this x column, this relation is a function.

Testing choice (2): When the relation is given as an equation, the simplest thing to do is to make a graph of that equation and see if it passes the vertical line test. The graph of a function will pass the "vertical line test," which means that there are no two points on the graph that have the same x-coordinate. If it is possible to draw a vertical line anywhere on the graph of a relation that will pass through more than one point, the relation will not be a function. The graph of the relation $3x + 2y = 4$ looks like this:

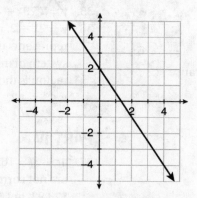

Since it passes the vertical line test, it is the graph of a function.

Testing choice (3): On this graph, there is no place where a vertical line will pass through more than one point, so this is the graph of a function.

Testing choice (4): When the relation is described with a mapping diagram, it is only a function if there is no place where the same element in the domain (the left oval) is being mapped to two different elements in the range (the right oval). Since in this mapping diagram the element 2 in the domain is mapped to both A and B, this relation is not a function.

The correct choice is **(4)**.

4. Graph both of the functions on the same set of axes. The graph for $g(x)$ is 3 units to the right of the graph of $f(x)$.

For the TI-84:

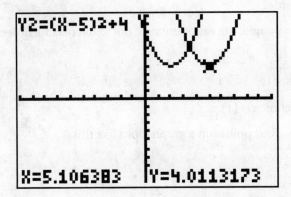

For the TI-Nspire:

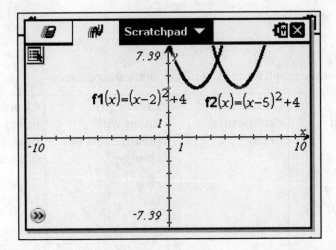

This question can also be done without a graphing calculator. An equation of the form $y = (x - h)^2 + k$ is called the *vertex form* of a parabola. The vertex will be (h, k).

The graph of function f will be a parabola with the vertex $(2, 4)$. The graph of the function g will be a parabola with the vertex $(5, 4)$. So, the graph of function g will be 3 units to the right of the graph of function f.

The correct choice is **(2)**.

5. When a polynomial is written in standard form, the term with the largest exponent will be first and each term after that will have a smaller exponent than the previous term. The coefficients before the variables are not relevant to this problem.

For Anne, the sequence of exponents is 7, 5, 3, 1, so the exponents are in decreasing order.

For the other people, the exponents are not in decreasing order.
Bob: 3, 5, 7, 1
Carrie: 1, 7, 5, 3
Dylan: 1, 3, 5, 7

The correct choice is **(1)**.

6. Plotting these four points on a graph looks like this:

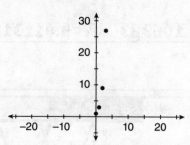

Here is what a graph looks like for each of the four choices:

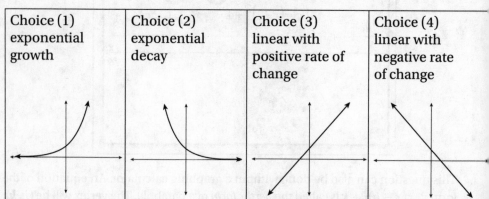

Choice (1) exponential growth	Choice (2) exponential decay	Choice (3) linear with positive rate of change	Choice (4) linear with negative rate of change

Of the four choices, the one that the four plotted points most resembles is choice (1).

The correct choice is **(1)**.

7. A rule of radicals is that $\sqrt{a} \cdot \sqrt{b} = \sqrt{ab}$. When this rule is applied to choice (1), it becomes $\sqrt{2} \cdot \sqrt{18} = \sqrt{2 \cdot 18} = \sqrt{36} = 6$, which is a rational number since it is a terminating decimal number. When an irrational number is expressed as a decimal, it will be nonterminating and nonrepeating.

The correct choice is **(1)**.

8. The x-intercepts of this graph are $(-2, 0)$, $(1, 0)$, and $(2, 0)$. If $(a, 0)$ is an x-intercept of the graph of a polynomial function, one of its factors will be $(x - a)$. So, for this function, $f(x) = (x - (-2))(x - 1)(x - 2) = (x + 2)(x - 1)(x - 2)$. Though this is not one of the choices, if you multiply the first and third factors together, $(x + 2)(x - 2) = x^2 - 4$. So, the function can also be written as $f(x) = (x - 1)(x^2 - 4)$.

Another way you could approach this question would be to graph the answer choices to see which one matches the given graph. When choice (3) is graphed on the graphing calculator, it looks like this:

For the TI-84:

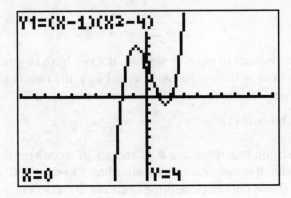

For the TI-Nspire:

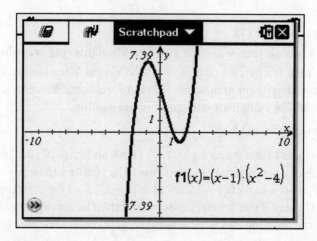

The correct choice is **(3)**.

9. She got from $p^2 + 5 = 8p - 7$ to $p^2 + 12 = 8p$ by adding 7 to both sides of the equation.

$$\begin{array}{r} p^2 + 5 = 8p - 7 \\ +7 = \quad +7 \\ \hline p^2 + 12 = 8p \end{array}$$

The addition property of equality states that if you have an equality and you add the same number to both sides of the equal sign, the new equation is also true. This is what was done here.

The correct choice is **(4)**.

10. The solution could be written as $x > 42$ and $x \leq 57$ since a person's height must meet both of those conditions. A shorthand for this "and" statement is $42 < x \leq 57$. This is <u>not</u> the same thing as choice (3), $42 < x$ or $x \leq 57$. The "or" statement is true if either of the conditions is met; so, for choice (3), someone under 42 inches would be allowed on the ride because his or her height is less than or equal to 57 inches.

The correct choice is **(1)**.

11. Make a sketch of the scenario in each of the answer choices and see which one looks most like a line.

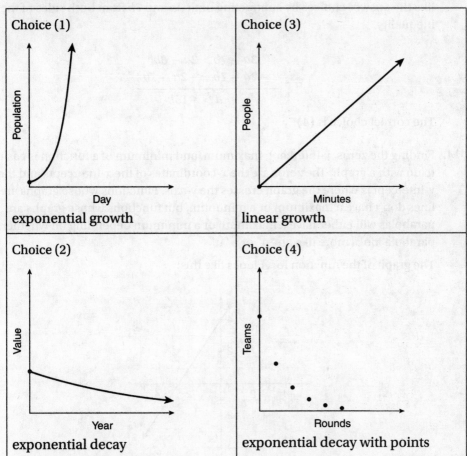

The correct choice is **(3)**.

12. A total of $42 + 26 = 68$ people preferred burgers. Of those 68 people, 26 of them were female. So, the percentage of people who preferred burgers who were female is $\frac{26}{68} \approx .382 = 38.2\%$.

The correct choice is **(2)**.

13. When the same number is added to or subtracted from both sides of an inequality, the inequality remains true. This inequality can be made to look like the answer choices by subtracting both $2a$ and $7b$ from both sides of the inequality.

$$\begin{array}{r} 3a + 7b > 2a - 8b \\ \underline{-2a - 7b = -2a - 7b} \\ a > -15b \end{array}$$

The correct choice is **(4)**.

14. Finding the zeros, y-intercept, maximum, and minimum of a function is easiest to do with a graph. The zeros are the x-coordinates of the x-intercepts and the y-intercept is where the graph crosses the y-axis. Functions whose graphs are lines don't have a maximum or a minimum, but functions whose graphs are parabolas will either have a maximum or a minimum depending on whether the parabola looks more like an "n" or a "u."

The graph of the function for A looks like this:

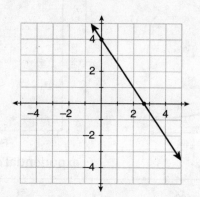

It has one zero at the x-intercept at $x = \dfrac{8}{3}$. The y-intercept is at $(0, 4)$. Because it is a line, it does not have a maximum or a minimum point.

The graph of the function for *B* looks like this:

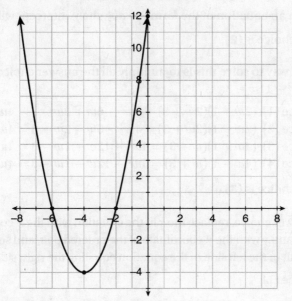

It has two zeros at the *x*-intercepts at $x = -2$ and $x = -6$. The *y*-intercept is at $(0, 12)$. Because it is a parabola shaped like a "u," it has a minimum point.

The graph of the function for *C* was given.

It has two zeros at the *x*-intercepts at $x = -2$ and $x = -4$. The *y*-intercept is at $(0, -8)$. Because it is a parabola shaped like an "n," it has a maximum point.

B has a minimum and *C* has a maximum.

The correct choice is **(3)**.

15. If Nicci's age is *a*, her sister's age is $2a - 7$ (<u>not</u> $7 - 2a$). The sum of their ages is $a + (2a - 7)$. Since the sum of their ages is 41, the equation is $a + (2a - 7) = 41$.

The correct choice is **(2)**.

16. Since the population is increasing each year, this is an example of exponential growth. For an equation of the form $y = ab^x$, exponential growth occurs when $b > 1$. This means that Roman numeral IV must be part of the solution.

Since in 2013 the population is 3810, $P(0) = 3810$. This fact can be used to solve for *a*.

$$P(0) = ab^0 = 3810$$
$$P(0) = a \cdot 1 = 3810$$
$$a = 3810$$

This means that Roman numeral I must be part of the solution.

In general, in an equation of the form $y = ab^x$, the y-intercept will be a value of a.

The correct choice is **(2)**.

17. The simplest way to solve this is to multiply all the answer choices to see which becomes $4w^2 - 11w - 3$.

Testing choice (1): $(2w + 1)(2w - 3) = 4w^2 - 6w + 2w - 3 = 4w^2 - 4w - 3$
Testing choice (2): $(2w - 1)(2w + 3) = 4w^2 + 6w - 2w - 3 = 4w^2 + 4w - 3$
Testing choice (3): $(4w + 1)(w - 3) = 4w^2 - 12w + 1w - 3 = 4w^2 - 11w - 3$
Testing choice (4): $(4w - 1)(w + 3) = 4w^2 + 12w - 1w - 3 = 4w^2 + 11w - 3$

The correct choice is **(3)**.

18. For each answer choice, replace the x in the equation with the x-coordinate of the ordered pair and y with the y-coordinate of the ordered pair, and see which one does not result in the left side of the equation equaling the right side of the equation.

Testing choice (1):

$$15.25 \overset{?}{=} 3(-1.5)^2 - (-1.5) + 7$$
$$15.25 \overset{?}{=} 3 \cdot 2.25 + 1.5 + 7$$
$$15.25 \overset{\checkmark}{=} 15.25$$

Testing choice (2):

$$7.25 \overset{?}{=} 3 \cdot 0.5^2 - 0.5 + 7$$
$$7.25 \overset{?}{=} 3 \cdot 0.25 - 0.5 + 7$$
$$7.25 \overset{\checkmark}{=} 7.25$$

Testing choice (3):

$$10.25 \overset{?}{=} 3 \cdot 1.25^2 - 1.25 + 7$$
$$10.25 \overset{?}{=} 3 \cdot 1.5625 - 1.25 + 7$$
$$10.25 \neq 10.4375$$

Testing choice (4):

$$23.25 \overset{?}{=} 3 \cdot 2.5^2 - 2.5 + 7$$
$$23.25 \overset{?}{=} 3 \cdot 6.25 - 2.5 + 7$$
$$23.25 \overset{\checkmark}{=} 23.25$$

The correct choice is **(3)**.

19. An arithmetic sequence is one in which each term is equal to the previous term plus or minus some constant.

Testing Roman numeral I: Since $4 = 2 + 2$, $6 = 4 + 2$, $8 = 6 + 2$, etc., this is an arithmetic sequence with a first term of 2 and a common difference of 2.

Testing Roman numeral II: Since $4 = 2 \cdot 2$, $8 = 4 \cdot 2$, $16 = 8 \cdot 2$, etc., this is <u>not</u> an arithmetic sequence. It is a geometric sequence.

Testing Roman numeral III: Since $a + 2 = a + 2$, $a + 4 = a + 2 + 2$, $a + 6 = a + 4 + 2$, etc., this is an arithmetic sequence with a first term of a and a common difference of 2.

The correct choice is **(2)**.

20. Choices (1) and (2) include negative numbers. Since you cannot purchase a negative amount of beef, you can eliminate these choices. Since beef does not have to be purchased in one-pound units (you can buy, for example, 1.7 pounds of beef), choice (3) can be eliminated also.

The correct choice is **(4)**.

21. Using the quadratic formula with $a = 1$, $b = -5$, and $c = -4$,

$$x = \frac{-b \pm \sqrt{b^2 - 4ac}}{2a} = \frac{-(-5) \pm \sqrt{(-5)^2 - 4(1)(-4)}}{2(1)}$$

$$= \frac{5 \pm \sqrt{25 + 16}}{2} = \frac{5 \pm \sqrt{41}}{2}$$

The correct choice is **(2)**.

22. Population standard deviation can be calculated with a graphing calculator.

For the TI-84:

Press [STAT] [1]. Clear all lists and enter the thirteen numbers into L1.

L1	L2	L3	1
80			
74			
76			
80			
80			
82			
82			

L1(13) =82

Press [STAT], [Right Arrow], [1] and select Calculate.

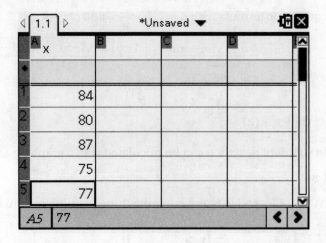

For the TI-Nspire:

From the home screen, select the Add Lists & Spreadsheet icon. Name column A "x" and fill in cells A1 to A13 with the thirteen numbers.

Press [menu] [4] [1] [1] for One-Variable Statistics.

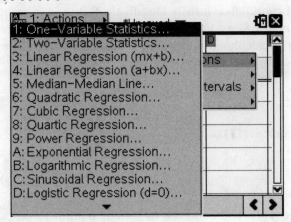

Select OK twice.

The population standard deviation is the one denoted with the symbol σx. Rounded to the nearest tenth, it is 3.5.

The correct choice is **(1)**.

23. In an exponential equation of the form $y = a(1 - r)^x$, r is the decay rate. For this question $1 - r = 0.98$, so $r = 0.02$.

The correct choice is **(2)**.

24. From the reference sheet provided on the Regents, 1 inch = 2.54 centimeters. To convert 91 centimeters into inches, divide 91 by 2.54 to get approximately 35.8 inches. Since 1 day equals 24 hours, the growth rate in inches per hour can be calculated by dividing the total growth by the total number of hours.

$$\frac{35.8}{24} \approx 1.49 \text{ inches per hour}$$

The correct choice is **(1)**.

Part II

Please Note: Some topics will no longer be tested on future Regents Algebra I exams. These are noted with an asterisk next to the specific questions.

25.

$$-\frac{2}{3}(x + 12) + \frac{2}{3}x = -\frac{5}{4}x + 2$$

$$-\frac{2}{3}x - 8 + \frac{2}{3}x = -\frac{5}{4}x + 2$$

$$-8 = -\frac{5}{4}x + 2$$

$$-2 = -2$$

$$-10 = -\frac{5}{4}x$$

$$-\frac{4}{5}(-10) = -\frac{4}{5}\left(-\frac{5}{4}x\right)$$

$$8 = x$$

26. $C = G - 3F = 3x^2 + 4 - 3(2x^2 + 6x - 5) = 3x^2 + 4 - 6x^2 - 18x + 15$
$= -3x^2 - 18x + 19$

27. A piecewise function is one that uses more than one rule, depending on what the input value to the function is. Make a table of values from $x = -5$ to $x = 6$. For the first seven values, use the top part of the piecewise function. For the last five values, use the bottom part of the piecewise function.

x	f(x)
−5	$\lvert -5 \rvert = 5$
−4	$\lvert -4 \rvert = 4$
−3	$\lvert -3 \rvert = 3$
−2	$\lvert -2 \rvert = 2$
−1	$\lvert -1 \rvert = 1$
0	$\lvert 0 \rvert = 0$
1	$\lvert 1 \rvert = 1$
2	$-2(2) + 10 = -4 + 10 = 6$
3	$-2(3) + 10 = -6 + 10 = 4$
4	$-2(4) + 10 = -8 + 10 = 2$
5	$-2(5) + 10 = -10 + 10 = 0$
6	$-2(6) + 10 = -12 + 10 = -2$

Plot these twelve points.

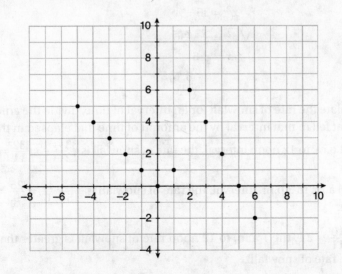

For $x = 2$, the bottom part of the piecewise function was used. For graphing, it is also useful to substitute $x = 2$ into the top part of the piecewise function because it is the boundary between the two pieces. When $x = 2$ is substituted into the top part of the piecewise function, the expression evaluates to 2. Plot the point $(2, 2)$ as an open circle since it is not an actual point on the graph. Then, connect the dots like this:

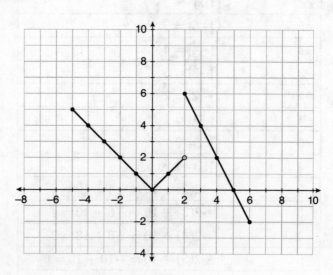

28.

$$\frac{5x^2}{5} = \frac{180}{5}$$

$$x^2 = 36$$

$$\sqrt{x^2} = \pm\sqrt{36}$$

$$x = \pm 6$$

$$\{-6, 6\}$$

29. To calculate the rate of snowfall for an interval of time, divide the amount of snow that fell in that interval by the amount of time that elapsed in that interval.

For the 1 a.m. to 12 noon interval, the rate of snowfall is $\frac{33 - 1}{12 - 1} = \frac{32}{11} = 2\frac{10}{11}$ inches per hour.

For the 6 a.m. to 3 p.m. interval, the rate of snowfall is $\frac{36 - 11}{15 - 6} = \frac{25}{9} = 2\frac{7}{9}$ inches per hour.

Since $2\frac{10}{11} > 2\frac{7}{9}$, the 1 a.m. to 12 noon rate of snowfall is greater than the 6 a.m. to 3 p.m. rate of snowfall.

30.

$$V = \frac{1}{3}\pi r^2 h$$

$$3V = 3 \cdot \frac{1}{3}\pi r^2 h$$

$$3V = \pi r^2 h$$

$$\frac{3V}{\pi r^2} = \frac{\pi r^2 h}{\pi r^2}$$

$$\frac{3V}{\pi r^2} = h$$

*31. A recursive formula describes how to get the next term of a sequence based on the previous terms. Substituting $n = 2, 3,$ and 4 into the recursive part of the equation, it becomes

$$a_2 = 2(a_{2-1} + 1) = 2(a_1 + 1) = 2(3 + 1) = 2 \cdot 4 = 8$$
$$a_3 = 2(a_{3-1} + 1) = 2(a_2 + 1) = 2(8 + 1) = 2 \cdot 9 = 18$$
$$a_4 = 2(a_{4-1} + 1) = 2(a_3 + 1) = 2(18 + 1) = 2 \cdot 19 = 38$$

32. The vertex form of a quadratic equation is $y = a(x - h)^2 + k$, where the vertex is (h, k). Completing the square is when you create a quadratic trinomial where the constant term is the square of half the coefficient on the x.

In the function $f(x) = x^2 - 2x - 8$, the coefficient on the x is -2. Calculate the square of half of this coefficient $\left(\frac{-2}{2}\right)^2 = (-1)^2 = +1$.

To complete the square, rewrite the -8 as $+ 1 - 9$. The function becomes $f(x) = x^2 - 2x + 1 - 9$. The first three terms are now a perfect square trinomial and can be factored.

$$f(x) = x^2 - 2x + 1 - 9 = (x - 1)^2 - 9$$

This is now in vertex form and the vertex is $(1, -9)$.

Part III

33. For Company A, x shirts cost $\$5x$ plus the $\$50$ set-up fee. The equation is $A = 5x + 50$. For Company B, x shirts cost $\$6x$ plus the $\$25$ set-up fee. The equation is $B = 6x + 25$.

For some values of x, A is more expensive and, for others, B is more expensive. To find the minimum number of shirts for it to be cheaper to use Company A, find the smallest integer that makes the inequality $5x + 50 < 6x + 25$ true.

*This topic will no longer be tested on future Regents Algebra I exams.

Solving the inequality becomes

$$5x + 50 < 6x + 25$$
$$-6x = -6x$$
$$-x + 50 < 25$$
$$-50 = -50$$
$$-x < -25$$
$$\frac{-x}{-1} > \frac{-25}{-1}$$
$$x > 25$$

The smallest integer that makes this true is 26. 25 is not the solution since it is not cheaper to buy 25 shirts from Company *A*; it is the same price as Company *B*.

34. The graph of function *f* is a parabola, and the graph of function *g* is a line. An important point to graph for a parabola is the vertex. In general, the *x*-coordinate of the vertex of a parabola is $x = -\dfrac{b}{2a} = -\dfrac{-8}{2 \cdot 2} = 2$. Choose at least three values smaller than 2 and three values greater than 2 for the *x* values in your chart.

x	f(x)	g(x)
−1	$2(-1)^2 - 8(-1) + 3 = 2 + 8 + 3 = 13$	$-2(-1) + 3 = 2 + 3 = 5$
0	$2(0)^2 - 8(0) + 3 = 0 + 0 + 3 = 3$	$-2(0) + 3 = 0 + 3 = 3$
1	$2(1)^2 - 8(1) + 3 = 2 - 8 + 3 = -3$	$-2(1) + 3 = -2 + 3 = 1$
2	$2(2)^2 - 8(2) + 3 = 8 - 16 + 3 = -5$	$-2(2) + 3 = -4 + 3 = -1$
3	$2(3)^2 - 8(3) + 3 = 18 - 24 + 3 = -3$	$-2(3) + 3 = -6 + 3 = -3$
4	$2(4)^2 - 8(4) + 3 = 32 - 32 + 3 = 3$	$-2(4) + 3 = -8 + 3 = -5$
5	$2(5)^2 - 8(5) + 3 = 50 - 40 + 3 = 13$	$-2(5) + 3 = -10 + 3 = -7$

The graph looks like this:

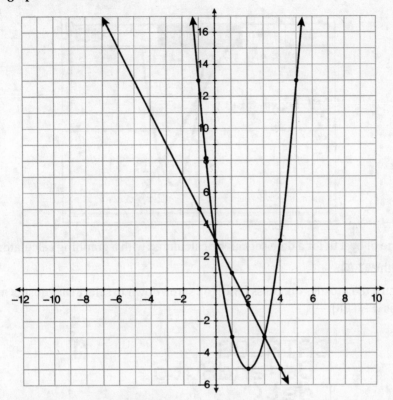

According to the chart, $f(0) = g(0) = 3$ and $f(3) = g(3) = -3$, so the two values of x for which $f(x) = g(x)$ are 0 and 3.

It can also be seen on the graph that the x-coordinates of the two points where the line intersects the parabola are 0 and 3.

These functions can also be graphed on a graphing calculator.

For the TI-84:

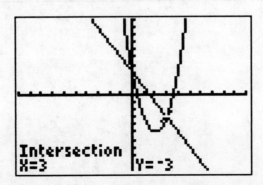

For the TI-Nspire:

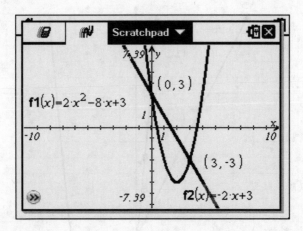

35. A line of best fit for a data set can be calculated with a graphing calculator.

For the TI-84:

To turn diagnostics on, press [2nd] [0], scroll down to select DiagnosticOn, and press [ENTER].

Press [STAT] [1] and enter the *x*-values into L1 and the *y*-values into L2.

L1	L2	L3	2
4	65		
4	67		
4	70		
6	82		
6	88		
7	82		
8	95		

L2(10) = 95

Press [STAT] [Right Arrow] [4] to get the LinReg function. Select Calculate.

LinReg
$y = ax + b$
$a = 7.792207792$
$b = 34.27272727$
$r^2 = .9634878259$
$r = .9815741571$

For the TI-Nspire:

From the home screen select the Add Lists & Spreadsheet icon. In the first row, label the A column "hours" and the B column "score." Enter the hours in cells A1 to A10. Enter the scores in cells B1 to B10.

	A hours	B score	C	D
◆				
1	0	35		
2	1	40		
3	2	46		
4	4	65		
5	4	67		

1.1 1.2 1.3 ▷ *Unsaved ▼

Move the cursor to cell C1. Press [menu] [4] [1] [3]. Set X List to "hours" and Y List to "score," and select OK.

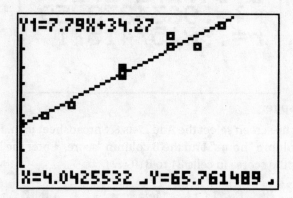

The line of best fit is $y = 7.79x + 34.27$.

This is the line of best fit graphed together with the scatterplot.

For the TI-84:

For the TI-Nspire:

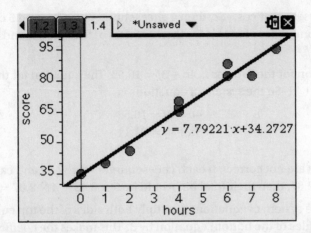

The correlation coefficient (r) is a measure of how close the points are to the line of best fit. If the line of best fit has a positive slope, the correlation coefficient will be positive. If the line of best fit has a negative slope, the correlation coefficient will be negative. The closer the correlation coefficient is to 1 (or to -1), the closer to the line of best fit the points will lie.

For this data set, the correlation coefficient is 0.98, which is very close to 1, so the data set is very nearly a line.

36. The line that is rising from bottom left to top right has a y-intercept of $(0, -2)$ and, since it passes through $(1, 0)$, it has a slope of 2. The equation of this line is $y = 2x - 2$. Since the area underneath that line is shaded, the first inequality is $y \leq 2x - 2$.

The line that is going down from top left to bottom right has a y-intercept of $(0, 3)$ and, since it passes through $(1, 0)$, it has a slope of -3. The equation of this line is $y = -3x + 3$. The second inequality is $y < -3x + 3$. There is a "$<$" instead of a "$\leq$."

So, the system of inequalities is

$$y \leq 2x - 2$$
$$y < -3x + 3$$

The region A represents all the ordered pairs that make up the solution set to the system of inequalities. Ordered pairs in this region will make both of the inequalities true.

The shaded gray region represents all the ordered pairs that make up the solution set to the inequality $y \leq 2x - 2$.

Part IV

37. If one cheeseburger costs \$$c$, then 4 cheeseburgers are \$$4c$ and 5 cheeseburgers are \$$5c$. If one order of fries is \$$f$, then 3 orders of fries are \$$3f$ and 4 orders of fries are \$$4f$.

The equation for the Browns is $4c + 3f = 16.53$. The equation for the Greens is $5c + 4f = 21.11$. So the system of equations is

$$4c + 3f = 16.53$$
$$5c + 4f = 21.11$$

The Greens are not correct. If each cheeseburger was \$2.49 and each order of fries was \$2.87, then their total bill would be $5 \times 2.49 + 4 \times 2.87 = 23.93 \neq 21.11$.

To solve the system of equations, multiply both sides of the top equation by -4 and both sides of the bottom equation by 3. This makes the coefficients of the f-variables opposites; so, when the two equations are added together, the f will be eliminated.

$$-16c - 12f = -66.12$$
$$15c + 12f = 63.33$$

Adding the two equations together becomes

$$-c = -2.79$$
$$c = 2.79$$

The price of one cheeseburger is \$2.79.

To find the price of an order of fries, substitute $c = 2.79$ into either of the original equations and solve for f.

$$4(2.79) + 3f = 16.53$$
$$11.16 + 3f = 16.53$$
$$-11.16 = -11.16$$
$$\frac{3f}{3} = \frac{5.37}{3}$$
$$f = 1.79$$

The price of one order of fries is \$1.79.

Topic	Question Numbers	Number of Points	Your Points	Your Percentage
1. Polynomials	5, 8, 26, 32	$2 + 2 + 2 + 2 = 8$		
2. Properties of Algebra	9, 25, 28, 30	$2 + 2 + 2 + 2 = 8$		
3. Functions	2, 3, 20, 29	$2 + 2 + 2 + 2 = 8$		
4. Creating and Interpreting Equations	11, 15, 23	$2 + 2 + 2 = 6$		
5. Inequalities	10, 13, 36	$2 + 2 + 4 = 8$		
6. Sequences and Series	19, 31	$2 + 2 = 4$		
7. Systems of Equations	33, 37	$4 + 6 = 10$		
8. Quadratic Equations and Factoring	1, 17, 21	$2 + 2 + 2 = 6$		
9. Regression	35	4		
10. Exponential Equations	6, 16	$2 + 2 = 4$		
11. Graphing	4, 14, 18, 27, 34	$2 + 2 + 2 + 2 + 4 = 12$		
12. Statistics	12, 22	$2 + 2 = 4$		
13. Number Properties	7	2		
14. Unit Conversions	24	2		

How to Convert Your Raw Score to Your Algebra I Regents Exam Score

The accompanying conversion chart must be used to determine your final score on the June 2019 Regents Exam in Algebra I. To find your final exam score, locate in the column labeled "Raw Score" the total number of points you scored out of a possible 86 points. Since partial credit is allowed in Parts II, III, and IV of the test, you may need to approximate the credit you would receive for a solution that is not completely correct. Then locate in the adjacent column to the right the scale score that corresponds to your raw score. The scale score is your final Algebra I Regents Exam score.

Regents Exam in Algebra I—June 2019
Chart for Converting Total Test Raw Scores
to Final Exam Scores (Scale Scores)

Raw Score	Scale Score	Performance Level	Raw Score	Scale Score	Performance Level	Raw Score	Scale Score	Performance Level
86	100	5	57	81	4	28	66	3
85	99	5	56	81	4	27	65	3
84	97	5	55	81	4	26	64	2
83	96	5	54	80	4	25	63	2
82	95	5	53	80	4	24	61	2
81	94	5	52	80	4	23	60	2
80	93	5	51	80	4	22	58	2
79	92	5	50	79	3	21	57	2
78	91	5	49	79	3	20	55	2
77	90	5	48	79	3	19	54	1
76	90	5	47	78	3	18	52	1
75	89	5	46	78	3	17	50	1
74	88	5	45	78	3	16	48	1
73	88	5	44	77	3	15	46	1
72	87	5	43	77	3	14	44	1
71	86	5	42	76	3	13	42	1
70	86	5	41	76	3	12	39	1
69	86	5	40	75	3	11	37	1
68	85	5	39	75	3	10	34	1
67	84	4	38	74	3	9	31	1
66	84	4	37	74	3	8	28	1
65	84	4	36	73	3	7	25	1
64	83	4	35	72	3	6	22	1
63	83	4	34	72	3	5	19	1
62	83	4	33	71	3	4	15	1
61	82	4	32	70	3	3	12	1
60	82	4	31	69	3	2	8	1
59	82	4	30	68	3	1	4	1
58	82	4	29	67	3	0	0	1

August 2019 Exam
Algebra I

High School Math Reference Sheet

Conversions

1 inch = 2.54 centimeters

1 meter = 39.37 inches

1 mile = 5280 feet

1 mile = 1760 yards

1 mile = 1.609 kilometers

1 kilometer = 0.62 mile

1 pound = 16 ounces

1 pound = 0.454 kilogram

1 kilogram = 2.2 pounds

1 ton = 2000 pounds

1 cup = 8 fluid ounces

1 pint = 2 cups

1 quart = 2 pints

1 gallon = 4 quarts

1 gallon = 3.785 liters

1 liter = 0.264 gallon

1 liter = 1000 cubic centimeters

Formulas

Triangle	$A = \frac{1}{2}bh$
Parallelogram	$A = bh$
Circle	$A = \pi r^2$
Circle	$C = \pi d$ or $C = 2\pi r$

Formulas (continued)

General Prisms	$V = Bh$
Cylinder	$V = \pi r^2 h$
Sphere	$V = \frac{4}{3}\pi r^3$
Cone	$V = \frac{1}{3}\pi r^2 h$
Pyramid	$V = \frac{1}{3}Bh$
Pythagorean Theorem	$a^2 + b^2 = c^2$
Quadratic Formula	$x = \dfrac{-b \pm \sqrt{b^2 - 4ac}}{2a}$
Arithmetic Sequence	$a_n = a_1 + (n-1)d$
Geometric Sequence	$a_n = a_1 r^{n-1}$
Geometric Series	$S_n = \dfrac{a_1 - a_1 r^n}{1 - r}$ where $r \neq 1$
Radians	1 radian = $\frac{180}{\pi}$ degrees
Degrees	1 degree = $\frac{\pi}{180}$ radians
Exponential Growth/Decay	$A = A_0 e^{k(t - t_0)} + B_0$

PART I

Answer all 24 questions in this part. Each correct answer will receive 2 credits. No partial credit will be allowed. For each statement or question, write in the space provided the numeral preceding the word or expression that best completes the statement or answers the question. [48 credits]

Please Note: Some topics will no longer be tested on future Regents Algebra I exams. These are noted with an asterisk next to the specific questions.

1. Bryan's hockey team is purchasing jerseys. The company charges $250 for a onetime set-up fee and $23 for each printed jersey. Which expression represents the total cost of x number of jerseys for the team?

   (1) $23x$

   (2) $23 + 250x$

   (3) $23x + 250$

   (4) $23(x + 250)$

   1 ___2___

2. Which table represents a function?

x	y
2	−3
3	0
4	−3
2	1

   (1)

x	y
−3	0
−2	1
−3	2
2	3

   (3)

x	y
1	2
1	3
1	4
1	5

   (2)

x	y
−2	−4
0	2
2	4
4	6

   (4)

   2 ___2___

3. Which expression is equivalent to $2(x^2 - 1) + 3x(x - 4)$?
$2x^2 - 2 + 3x^2 - 12x$

(1) $5x^2 - 5$ (3) $5x^2 - 12x - 1$

(2) $5x^2 - 6$ (4) $5x^2 - 12x - 2$ 3

4. The value of x that satisfies the equation $\frac{4}{3} = \frac{x + 10}{15}$ is

(1) -6 ✗   (3) 10

(2) 5 ✗   (4) 30 4

5. Josh graphed the function $f(x) = -3(x - 1)^2 + 2$. He then graphed the function $g(x) = -3(x - 1)^2 - 5$ on the same coordinate plane. The vertex of $g(x)$ is

(1) 7 units below the vertex of $f(x)$

(2) 7 units above the vertex of $f(x)$

(3) 7 units to the right of the vertex of $f(x)$

(4) 7 units to the left of the vertex of $f(x)$ 5

6. A survey was given to 12th-grade students of West High School to determine the location for the senior class trip. The results are shown in the table below.

	Niagara Falls	Darien Lake	New York City
Boys	56	74	103
Girls	71	92	88

To the *nearest percent*, what percent of the boys chose Niagara Falls?

(1) 12   (3) 44

(2) 24   (4) 56   6 _____

7. Which type of function is shown in the graph below?

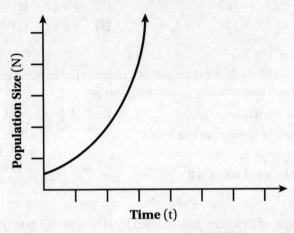

(1) linear   (3) square root
(②) exponential   (4) absolute value   7 _2_

8. The expression $16x^2 - 81$ is equivalent to

(1) $(8x - 9)(8x + 9)$ (3) $(4x - 9)(4x + 9)$
(2) $(8x - 9)(8x - 9)$ (4) $(4x - 9)(4x - 9)$ 8 _2_

9. The owner of a landscaping business wants to know how much time, on average, his workers spend mowing one lawn. Which is the most appropriate rate with which to calculate an answer to his question?

(1) lawns per employee   (3) employee per lawns
(2) lawns per day   (④) hours per lawn   9 _2_

10. A ball is thrown into the air from the top of a building. The height, $h(t)$, of the ball above the ground t seconds after it is thrown can be modeled by $h(t) = -16t^2 + 64t + 80$. How many seconds after being thrown will the ball hit the ground?

(①) 5   (3) 80
(2) 2   (4) 144   10 _2_

Handwritten at top of page: $4 + 18 + 144 = x^2 + 24x + 144$
$4 + 18 = x^2 + 24x$

11. Which equation is equivalent to $y = x^2 + 24x - 18$?

    (1) $y = (x + 12)^2 - 162$ (3) $y = (x - 12)^2 - 162$

    (2) $y = (x + 12)^2 + 126$ (4) $y = (x - 12)^2 + 126$ 11 _____

12. When $(x)(x - 5)(2x + 3)$ is expressed as a polynomial in standard form, which statement about the resulting polynomial is true?

    (1) The constant term is 2.

    (2) The leading coefficient is 2.

    (3) The degree is 2.

    (4) The number of terms is 2.   12 _____

13. The population of a city can be modeled by $P(t) = 3810(1.0005)^{7t}$, where $P(t)$ is the population after t years. Which function is approximately equivalent to $P(t)$?

    (1) $P(t) = 3810(0.1427)^t$

    (2) $P(t) = 3810(1.0035)^t$

    (3) $P(t) = 26{,}670(0.1427)^t$

    (4) $P(t) = 26{,}670(1.0035)^t$ 13 _____

14. The functions $f(x)$ and $g(x)$ are graphed on the set of axes below.

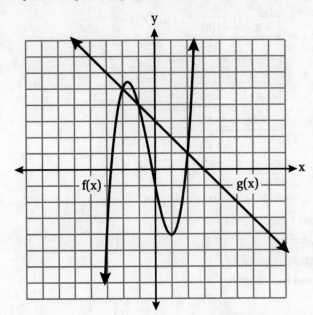

For which value of x is $f(x) \neq g(x)$?

(1) −1   (3) 3

(2) 2   (4) −2   14 __2__

15. What is the range of the box plot shown below?

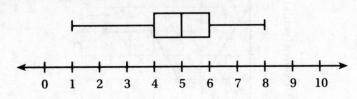

(1) 7   (3) 3

(2) 2   (4) 4   15 __2__

16. Which expression is *not* equivalent to $2x^2 + 10x + 12$?

(1) $(2x + 4)(x + 3)$ (3) $(2x + 3)(x + 4)$

(2) $(2x + 6)(x + 2)$ (4) $2(x + 3)(x + 2)$ 16 __2__

$2x^2 \neq 6x + 9x + 12$

17. The quadratic functions $r(x)$ and $q(x)$ are given below.

x	r(x)
−4	−12
−3	−15
−2	−16
−1	−15
0	−12
1	7

$q(x) = x^2 + 2x - 8$

The function with the *smallest* minimum value is

(1) $q(x)$, and the value is −9
(2) $q(x)$, and the value is −1
(3) $r(x)$, and the value is −16
(4) $r(x)$, and the value is −2

17 _____

18. A child is playing outside. The graph below shows the child's distance, $d(t)$, in yards from home over a period of time, t, in seconds.

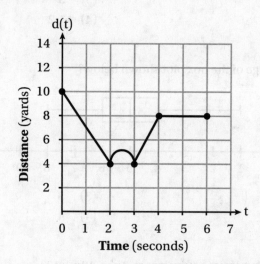

Which interval represents the child constantly moving closer to home?

(1) $0 \leq t \leq 2$ (3) $3 \leq t \leq 4$
(2) $2 \leq t \leq 3$ (4) $4 \leq t \leq 6$

18 _____

*19. If $a_1 = 6$ and $a_n = 3 + 2(a_{n-1})^2$, then a_2 equals

(1)   75

(2)   147 $a_2 = 3 + 2(a_2 - 1)^2$

(3)   180

(4)   900

19 _____

$a_2 = 3 + 2(a_6)^2$ $a_2 =$

20. The length of a rectangular patio is 7 feet more than its width, w. The area of a patio, $A(w)$, can be represented by the function

(1) $A(w) = w + 7$

(2) $A(w) = w^2 + 7w$

(3) $A(w) = 4w + 14$

(4) $A(w) = 4w^2 + 28w$ 20 _____

21. A dolphin jumps out of the water and then back into the water. His jump could be graphed on a set of axes where x represents time and y represents distance above or below sea level. The domain for this graph is best represented using a set of

(1)   integers

(2)   positive integers

(3)   real numbers

(4)   positive real numbers   21 _____

*22. Which system of linear equations has the same solution as the one shown below?

$2 - 12$

$$x - 4y = -10$$
$$x + y = 5$$

$2 \quad 3$

(1) $5x = 10$
      $x + y = 5$

(2) $-5y = -5$
      $x + y = 5$

(3) $-3x = -30$
      $x + y = 5$

(4) $-5y = -5$
      $x - 4y = -10$ 22 _____

---

*This topic will no longer be tested on future Regents Algebra I exams.

23. Which interval represents the range of the function $h(x) = 2x^2 - 2x - 4$?

   (1) $(0.5, \infty)$ (3) $[0.5, \infty)$

   (2) $(-4.5, \infty)$ (4) $[-4.5, \infty)$ 23 _____

24. What is a common ratio of the geometric sequence whose first term is 5 and third term is 245?

   (1) 7   (3) 120

   (2) 49   (4) 240   24 _____

PART II

Answer all 8 questions in this part. Each correct answer will receive 2 credits. Clearly indicate the necessary steps, including appropriate formula substitutions, diagrams, graphs, charts, etc. For all questions in this part, a correct numerical answer with no work shown will receive only 1 credit. [16 credits]

25. If $g(x) = -4x^2 - 3x + 2$, determine $g(-2)$.

$$g(-2) = -4(-2)^2 - 3(-2) + 2$$

$$g(-2) = 64 + 6 + 2$$

$$g(-2) = 72$$

26. A student is in the process of solving an equation. The original equation and the first step are shown below.

Original: $3a + 6 = 2 - 5a + 7$
Step one: $3a + 6 = 2 + 7 - 5a$

Which property did the student use for the first step? Explain why this property is correct.

commuative property

they just switched the order around

2

27. On the set of axes below, graph the line whose equation is $2y = -3x - 2$.

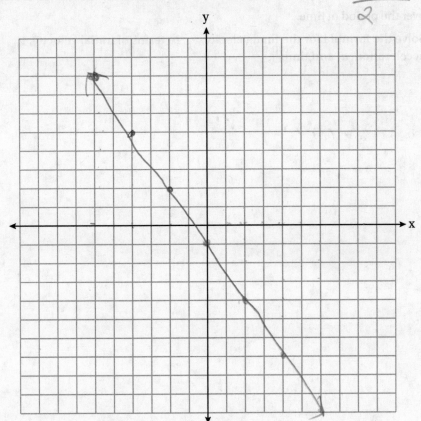

This linear equation contains the point $(2, k)$. State the value of k.

−4

2

28. The formula $a = \dfrac{v_f - v_i}{t}$ is used to calculate acceleration as the change in velocity over the period of time.

Solve the formula for the final velocity, v_f, in terms of initial velocity, v_i, acceleration, a, and time, t.

$$a \cdot t + v_i = v_f$$

2

29. Solve $\frac{3}{5}x + \frac{1}{3} < \frac{4}{5}x - \frac{1}{3}$ for x.

$$\frac{1}{3} < \frac{1}{5}x - \frac{1}{3}$$

$$\frac{5}{3} < x - \frac{1}{3}$$

$$\frac{6}{3} < x$$

$$2 < x$$

30. Is the product of two irrational numbers always irrational? Justify your answer.

yes.

irrational + irrational always
= irrational

31. Solve $6x^2 - 42 = 0$ for the exact values of x.

$$6x^2 = 42$$

$$6x^2 = 36$$

$$x^2 = \sqrt{36}$$

$$x = 6$$

32. Graph the function: $h(x) = \begin{cases} 2x - 3, & x < 0 \\ x^2 - 4x - 5, & 0 \le x \le 5 \end{cases}$

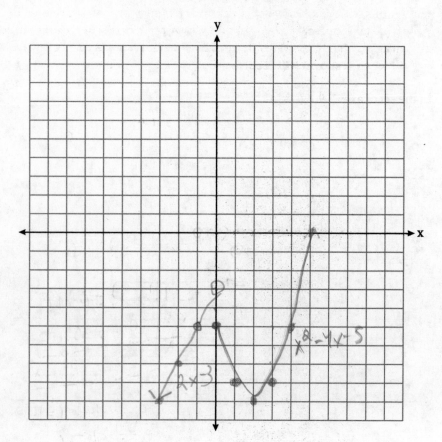

PART III

Answer all 4 questions in this part. Each correct answer will receive 4 credits.
Clearly indicate the necessary steps, including appropriate formula substitutions,
diagrams, graphs, charts, etc. For all questions in this part, a correct numerical
answer with no work shown will receive only 1 credit.   [16 credits]

33. On the set of axes below, graph the following system of inequalities:

$$2x + y \geq 8$$
$$y - 5 < 3x$$

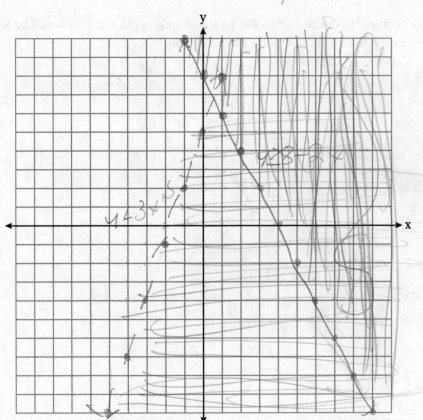

Determine if the point (1, 8) is in the solution set. Explain your answer.

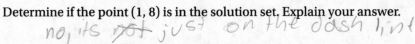

34. On the day Alexander was born, his father invested $5000 in an account with a 1.2% annual growth rate. Write a function, $A(t)$, that represents the value of this investment t years after Alexander's birth.

$$f(x) = \underset{a}{\tfrac{1}{t}}(1+r)^x$$

$$f(x) = 5000(1+.0012)^x$$

Determine, to the *nearest dollar*, how much more the investment will be worth when Alexander turns 32 than when he turns 17.

$$f(x) = 5000(1+.0012)^{17}$$

$$f(x) = 5000(1.0012)^{17}$$

$$f(x) = 5000(1.02059702)$$

$$f(x) = 5103$$

$$f(x) = 5000(1.012)^{32}$$

$$f(x) = 5000(1.464793487)$$

$$f(x) = 7324$$

$$\begin{array}{r} 7324 \\ -\ 5103 \\ \hline 2221 \end{array}$$

35. Stephen collected data from a travel website. The data included a hotel's distance from Times Square in Manhattan and the cost of a room for one weekend night in August. A table containing these data appears below.

Distance From Times Square (city blocks) (x)	0	0	1	1	3	4	7	11	14	19
Cost of a Room (dollars) (y)	293	263	244	224	185	170	219	153	136	111

Write the linear regression equation for this data set. Round all values to the *nearest hundredth*.

$$y = -7.76x + 246.34$$

State the correlation coefficient for this data set, to the *nearest hundredth*.

$-.88$

Explain what the sign of the correlation coefficient suggests in the context of the problem.

a strong negative correlation

36. A snowstorm started at midnight. For the first 4 hours, it snowed at an average rate of one-half inch per hour.

The snow then started to fall at an average rate of one inch per hour for the next 6 hours.

Then it stopped snowing for 3 hours.

Then it started snowing again at an average rate of one-half inch per hour for the next 4 hours until the storm was over.

On the set of axes below, graph the amount of snow accumulated over the time interval of the storm.

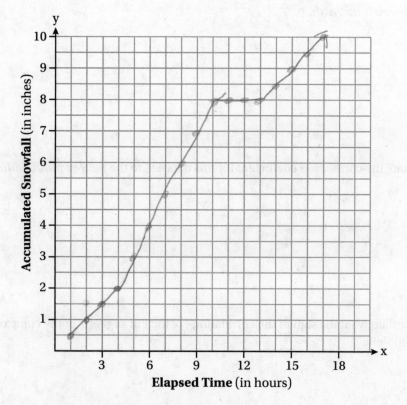

Determine the average rate of snowfall over the length of the storm. State the rate, to the *nearest hundredth of an inch per hour*.

.56 inches per hour

3

PART IV

Answer the question in this part. A correct answer will receive 6 credits. Clearly indicate the necessary steps, including appropriate formula substitutions, diagrams, graphs, charts, etc. A correct numerical answer with no work shown will receive only 1 credit. [6 credits]

37. Allysa spent $35 to purchase 12 chickens. She bought two different types of chickens. Americana chickens cost $3.75 each and Delaware chickens cost $2.50 each.

Write a system of equations that can be used to determine the number of Americana chickens, A, and the number of Delaware chickens, D, she purchased.

$$3.75A + 2.50D = 35$$
$$A + D = 12$$

Determine algebraically how many of each type of chicken Allysa purchased.

$$3.75A + 2.50D = 35$$
$$-2.50A - 2.50D = -30$$
$$\overline{}$$
$$1.25A = 5$$
$$A = 4$$

$$4 + 8 = 12$$

Each Americana chicken lays 2 eggs per day and each Delaware chicken lays 1 egg per day. Allysa only sells eggs by the full dozen for $2.50. Determine how much money she expects to take in at the end of the first week with her 12 chickens.

$$8 + 8 = 16$$

$$7 \cdot 16 = 112$$

$$\frac{112}{12} \approx 9$$

$$\$22.50$$

6

Answers August 2019

Algebra I

Answer Key

PART I

1. (3)	5. (1)	9. (4)	13. (2)	17. (3)	21. (4)
2. (4)	6. (2)	10. (1)	14. (3)	18. (1)	22. (1)
3. (4)	7. (2)	11. (1)	15. (1)	19. (1)	23. (4)
4. (3)	8. (3)	12. (2)	16. (3)	20. (2)	24. (1)

PART II

25. -8

26. Commutative property of addition

27. 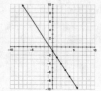 $k = -4$

28. $v_f = at + v_i$

29. $x > \dfrac{10}{3}$

30. No. For example, $\sqrt{2} \cdot \sqrt{2} = 2$.

31. $x = \pm\sqrt{7}$

32.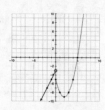

PART III

33. $(1, 8)$ is not part of the solution set.

34. $A(t) = 5000\,(1.012)^t$, $1200

35. $y = -7.76x + 246.34$, $r = -0.88$, which is negative because the line of best fit has a negative slope.

36.    0.59 inches per hour

PART IV

37. $A + D = 12$
    $3.75A + 2.50D = 35$
    $A = 4, D = 8$, $22.50

In **Parts II–IV**, you are required to show how you arrived at your answers. For sample methods of solutions, see the *Answer Explanations* section.

Answer Explanations

Part I

Please Note: Some topics will no longer be tested on future Regents Algebra I exams. These are noted with an asterisk next to the specific questions.

1. Since each printed jersey costs 23 dollars, x jerseys cost $23x$ dollars. When the cost of the jerseys is added to the one-time setup fee of 250 dollars, the total cost is then $23x + 250$ dollars.

   In general, when an equation of the form $mx + b$ models a real-world scenario, the b represents the fixed cost and the m represents the increase in the total for each time x increases by 1.

   The correct choice is **(3)**.

2. When a function is represented in a table, there will be no repeated values in the x column. In choice (1), there are two 2's. In choice (2), there are four 1's. In choice (3), there are two -3's. In choice (4), there are no repeated values in the x column, so this table represents a function.

   Even if in choice (4) there were repeated values among the y-values, it would still be a function. All that matters is that there are no repeated values among the x-values.

   The correct choice is **(4)**.

3. Use the distributive property and then combine like terms.

$$2(x^2 - 1) + 3x(x - 4) = 2x^2 - 2 + 3x^2 - 12x$$
$$= 5x^2 - 12x - 2$$

   The correct choice is **(4)**.

4. Cross multiply and then solve for x.

$$\frac{4}{3} = \frac{x + 10}{15}$$
$$4 \cdot 15 = 3(x + 10)$$
$$60 = 3x + 30$$
$$-30 = -30$$
$$30 = 3x$$
$$\frac{30}{3} = \frac{3x}{3}$$
$$10 = x$$

Another way to solve this is to make the two fractions have a common denominator of 15 and then use the fact that the numerators will then be equal.

$$\frac{4}{3} = \frac{x + 10}{15}$$
$$\frac{20}{15} = \frac{x + 10}{15}$$
$$20 = x + 10$$
$$-10 = -10$$
$$10 = x$$

The correct choice is **(3)**.

5. Graph both functions on your graphing calculator.

On the TI-84:

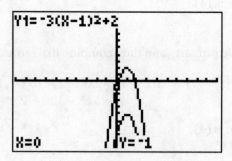

On the TI-Nspire:

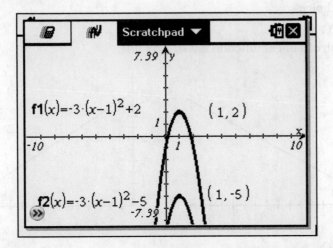

In a parabola that opens downward, the vertex is the maximum point. The vertex of the graph of function f is $(1, 2)$. The vertex of the graph of function g is $(1, -5)$. So, the vertex of the graph of $g(x)$ is 7 units below the vertex of function $f(x)$.

This can also be done without a graphing calculator by using the properties of the vertex form of an equation for a parabola.

When a function is in the form $f(x) = a(x - h)^2 + k$, the graph will be a parabola with its vertex at (h, k). Function f is in exactly this form, so its vertex is $(1, 2)$. Function g can be expressed as $g(x) = -3(x - 1)^2 + (-5)$, so its vertex is $(1, -5)$. The point $(1, -5)$ is 7 units below the point $(1, 2)$.

The correct choice is **(1)**.

6. The percent of boys who chose Niagara Falls can be calculated by dividing the number of boys who chose Niagara Falls by the total number of boys. First find the total number of boys.

$$56 + 74 + 103 = 233$$

Now find the percent of boys.

$$\frac{56}{233} \approx 0.24 = 24\%$$

The correct choice is **(2)**.

7. Compare the shape of this graph to the shape of typical graphs for each of the answer choices.

Choice (1):

The graph of a linear function is a straight line.

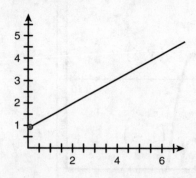 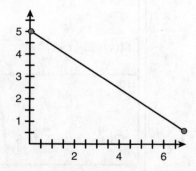

Choice (2):

The graph of an exponential function looks like a playground slide.

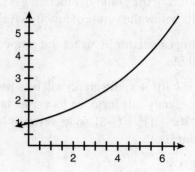 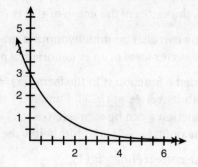

Choice (3):

The graph of a square root function looks like half of a sideways parabola.

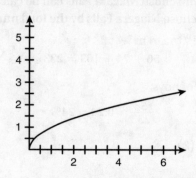

Choice (4):

The graph of an absolute value function looks like the letter "V."

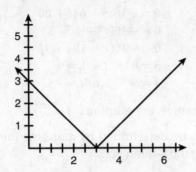

The given graph most resembles a portion of the graph of an exponential function.

The correct choice is **(2)**.

8. The difference of perfect squares factoring pattern is $a^2 - b^2 = (a - b)(a + b)$. Because both terms of the given expression are perfect squares, the expression can be written as the difference of perfect squares.

$$16x^2 - 81 = (4x)^2 - 9^2$$
$$= (4x - 9)(4x + 9)$$

The correct choice is **(3)**.

9. Depending on what the owner of the business is looking to measure, any of these choices could be appropriate. Since in this problem the owner wants to know the average amount of time his team spends mowing one lawn, he would calculate the answer by dividing the total number of hours spent by the total number of lawns mowed. For example, if the workers took 8 hours to mow 4 lawns, the average amount of time to mow one lawn would be $\dfrac{8 \text{ hours}}{4 \text{ lawns}} = \dfrac{8}{4} = 2$ hours per lawn.

The correct choice is **(4)**.

10. When the ball hits the ground, the height of the ball above the ground will be 0. To find when this happens, solve the equation $0 = -16t^2 + 64t + 80$.

$$0 = -16t^2 + 64t + 80$$
$$0 = -16(t^2 - 4t - 5)$$
$$0 = -16(t - 5)(t + 1)$$
$$t - 5 = 0 \qquad t + 1 = 0$$
$$t = 5 \qquad \text{or } t = -1$$

Since time must be positive, the solution is $t = 5$.

This answer can also be determined by graphing the function on a graphing calculator and finding the x-intercept.

On the TI-84:

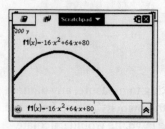 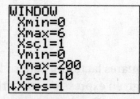 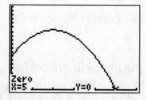

On the TI-Nspire:

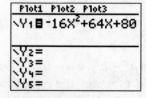  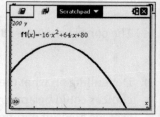

The correct choice is **(1)**.

11. The given equation can be expressed in a form like the answer choices by using the completing the square process.

The first step is to eliminate the coefficient from the right-hand side by adding 18 to both sides of the equation.

$$y = x^2 + 24x - 18$$
$$+18 = +18$$
$$y + 18 = x^2 + 24x$$

Next, square half the coefficient of the x-term and add that to both sides of the equation.

$$\left(\frac{24}{2}\right)^2 = 12^2 = 144$$
$$y + 18 + 144 = x^2 + 24x + 144$$
$$y + 162 = x^2 + 24x + 144$$

The right-hand side of the equation is now a perfect square trinomial and can be factored.

$$y + 162 = x^2 + 24x + 144$$
$$y + 162 = (x + 12)^2$$

Isolate the y by subtracting 162 from both sides of the equation.

$$y + 162 = (x + 12)^2$$
$$-162 = -162$$
$$y = (x + 12)^2 - 162$$

This question can also be solved without completing the square by just simplifying each of the four answer choices to see which is equivalent to $y = x^2 + 24x - 18$.

Testing choice (1):

$$y = (x + 12)^2 - 162$$
$$y = x^2 + 24x + 144 - 162$$
$$y \overset{\checkmark}{=} x^2 + 24x - 18$$

Testing choice (2):

$$y = (x + 12)^2 + 126$$
$$y = x^2 + 24x + 144 + 126$$
$$y = x^2 + 24x + 270$$
$$y \neq x^2 + 24x - 18$$

Testing choice (3):

$$y = (x - 12)^2 - 162$$
$$y = x^2 - 24x + 144 - 162$$
$$y = x^2 - 24x - 18$$
$$y \neq x^2 + 24x - 18$$

Testing choice (4):

$$y = (x - 12)^2 + 126$$
$$y = x^2 - 24x + 144 + 126$$
$$y = x^2 - 24x + 270$$
$$y \neq x^2 + 24x - 18$$

The correct choice is **(1)**.

12. First multiply the factors together.

$$(x)(x - 5)(2x + 3) = (x^2 - 5x)(2x + 3)$$
$$= 2x^3 + 3x^2 - 10x^2 - 15x$$
$$= 2x^3 - 7x^2 - 15x$$

This polynomial has three terms: $2x^3$, $7x^2$, and $15x$.

The constant term is 0 since there is no term that is not multiplied by x or a power of x. Because the highest exponent is a 3, the degree of this polynomial is 3.

The leading coefficient is the number that is multiplied by the highest power of x. Since the term $2x^3$ has the highest power of x, the leading coefficient is 2.

The correct choice is **(2)**.

13. One of the laws of exponents is $x^{ab} = (x^a)^b$. Apply this to the given expression.

$$P(t) = 3810(1.0005)^{7t}$$
$$= 3810(1.0005^7)^t$$
$$= 3810(1.0035)^t$$

Another way to solve this problem is to pick any value you like for t, such as $t = 10$. Then calculate $P(10)$ in the original function.

$$P(10) = 3810(1.0005)^{7 \cdot 10}$$
$$= 3810(1.0005)^{70}$$
$$= 3810(1.03561065)$$
$$\approx 3945.7$$

Then test for $t = 10$ in each of the answer choices. When you test choice (2), you find approximately the same result.

$$P(10) = 3810(1.0035)^{10}$$
$$= 3810(1.035556427)$$
$$\approx 3945.5$$

The correct choice is **(2)**.

14. The x-coordinates of the intersection points of the line and the curve are the solutions to $f(x) = g(x)$. Since the line intersects the curve at $(-2, 5)$, $(-1, 4)$, and $(2, 1)$, the three x-values that make $f(x) = g(x)$ are -2, -1, and 2. Of the four choices, each of these three values appear. However, 3 is not one of the solutions.

    The correct choice is **(3)**.

15. In a box plot, the left endpoint represents the minimum value. For this problem, the minimum value is 1. The right endpoint represents the maximum value. For this problem, the maximum value is 8. The range is the difference between the maximum value and the minimum value, which is $8 - 1 = 7$.

    The correct choice is **(1)**.

16. Factor the given expression by first factoring out the greatest common factor of 2.
    $$2x^2 + 10x + 12 = 2(x^2 + 5x + 6)$$
    $$= 2(x + 3)(x + 2)$$

    Since this is choice (4) and the question is asking which choice is *not* equivalent, choice (4) can be eliminated.

    To find the other two equivalent expressions, distribute the 2 to each of the factors.
    $$2(x + 3)(x + 2) = (2x + 6)(x + 2)$$

    So, choice (2) can be eliminated.
    $$2(x + 3)(x + 2) = (x + 3)2(x + 2)$$
    $$= (x + 3)(2x + 4)$$

    So, choice (1) can be eliminated.

    Another way to solve this problem is to multiply out each of the four choices to see which one is not equivalent to the given expression.

    Testing choice (1):
    $$(2x + 4)(x + 3) = 2x^2 + 6x + 4x + 12$$
    $$= 2x^2 + 10x + 12$$

    Testing choice (2):
    $$(2x + 6)(x + 2) = 2x^2 + 4x + 6x + 12$$
    $$= 2x^2 + 10x + 12$$

Testing choice (3):

$$(2x + 3)(x + 4) = 2x^2 + 8x + 3x + 12$$
$$= 2x^2 + 11x + 12$$
$$\neq 2x^2 + 10x + 12$$

Testing choice (4):

$$2(x + 3)(x + 2) = (2x + 6)(x + 2)$$
$$= 2x^2 + 4x + 6x + 12$$
$$= 2x^2 + 10x + 12$$

The correct choice is **(3)**.

17. Based on the data in the table, $r(-2) = -16$ is the minimum of function r. For function q, you can graph the function on a graphing calculator and find the coordinates of the minimum point using the minimum function.

    On the TI-84:

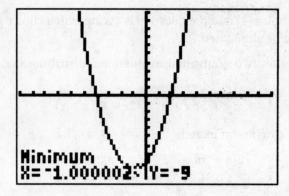

On the TI-Nspire:

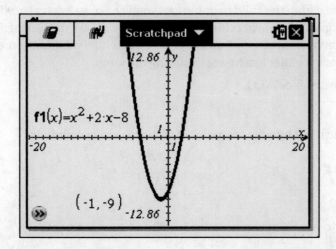

The minimum point is located at $(-1, -9)$, so the minimum value for function q is -9.

Note that the x-coordinate is not the minimum; the y-coordinate is the minimum.

Even without using a graphing calculator, you can calculate the x-coordinate of the minimum of function q with the equation $x = -\dfrac{b}{2a}$.

$$x = -\frac{b}{2a}$$

$$= -\frac{2}{2 \cdot 1}$$

$$= -1$$

Then calculate the value of q when $x = -1$.

$$q(-1) = (-1)^2 + 2(-1) - 8$$
$$= 1 - 2 - 8$$
$$= -9$$

Since $-16 < -9$, function r has the smallest minimum value.

The correct choice is (**3**).

18. At $t = 0$, the child is 10 yards away from home. As the child gets closer to home, the distance from the child to home gets smaller. So, an interval in which the child is constantly moving closer to home will be always decreasing. This happens in the interval between $t = 0$ and $t = 2$. The other intervals have some parts where the distance is either increasing or staying the same.

The correct choice is **(1)**.

*19. Substitute $n = 2$ into the second part of the recursive definition.

$$
\begin{aligned}
a_2 &= 3 + 2(a_{2-1})^2 \\
&= 3 + 2(a_1)^2 \\
&= 3 + 2 \cdot 6^2 \\
&= 3 + 2 \cdot 36 \\
&= 3 + 72 \\
&= 75
\end{aligned}
$$

The correct choice is **(1)**.

20. If the width is w feet and the length is 7 feet more than the width of a rectangular patio, the length can be expressed as $w + 7$ feet. The area of a rectangle is length times width.

$$
\begin{aligned}
A(w) &= lw \\
&= (w + 7)w \\
&= w^2 + 7w
\end{aligned}
$$

The correct choice is **(2)**.

21. A sketch of the graph could look like the following.

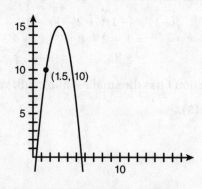

The domain is the set of x-coordinates of all the points on the graph. This domain is not limited to integers since there are points that have x-coordinates that are not integers, such as $(1.5, 10)$. Numbers between integers qualify as real numbers. Since the graph starts at $x = 0$, there are also no negative x-values. Negative x-values would be used for moments before the dolphin begins jumping. So, positive real numbers would be most appropriate for the domain of this graph.

If the question instead asked for the range of the graph, then real numbers, including negative values, would be appropriate since there are times where the dolphin is under water. The y-coordinates of the points representing those underwater moments would be negative.

The correct choice is **(4)**.

22. Solve the given system of equations with the elimination method. First multiply both sides of the second equation by -1 and then add the first equation to that result.

$$x - 4y = -10$$
$$-1(x + y) = -1(5)$$

$$
\begin{aligned}
& x - 4y = -10 \\
+ \;\; & -x - y = -5 \\
\hline
& \frac{-5y}{-5} = \frac{-15}{-5} \\
& y = 3
\end{aligned}
$$

Substitute $y = 3$ into either of the two original equations to solve for x.

$$
\begin{aligned}
x + 3 &= 5 \\
-3 &= -3 \\
x &= 2
\end{aligned}
$$

The solution to the given system is $(2, 3)$.

The answer choices are much simpler systems since in each of them, one of the equations has just one variable. Also notice that the other equation in each choice matches one of the equations in the given system. So, all that needs to be done is to check each choice to see if the solution to the one-variable equation matches either $x = 2$ or $y = 3$.

This topic will no longer be tested on future Regents Algebra I exams.

Checking choice (1):

$$\frac{5x}{5} = \frac{10}{5}$$

$$x = 2$$

Checking choice (2):

$$\frac{-5y}{-5} = \frac{-5}{-5}$$

$$y = 1$$

Checking choice (3):

$$\frac{-3x}{-3} = \frac{-30}{-3}$$

$$x = 10$$

Checking choice (4):

$$\frac{-5y}{-5} = \frac{-5}{-5}$$

$$y = 1$$

The correct choice is **(1)**.

23. The range is the set of y-coordinates of all the points on the graph of a function. Graph this function on a graphing calculator.

On the TI-84:

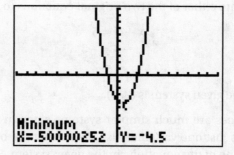

On the TI-Nspire:

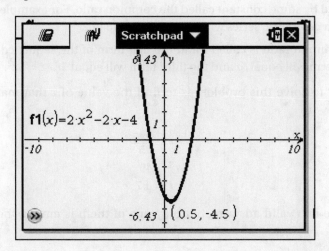

The lowest y-coordinate of the graph is -4.5. There is no highest y-coordinate since both branches of the parabola go up forever.

One way to express this range is $y \geq -4.5$. The answer choices, however, use interval notation.

The only difference between choices (2) and (4) is that choice (2) has a parenthesis "(" on the left side of the interval and choice (4) has a square bracket "[."

The square bracket "[" is used when the interval does contain the first number. In this case, since the minimum point of the parabola is -4.5, the square bracket "[" is the appropriate notation.

The correct choice is **(4)**.

24. A geometric sequence is one in which each term is equal to the previous term multiplied by some constant called the common ratio. For example, 3, 6, 12, 24, . . . is a geometric series with a common ratio of 2.

If the common ratio is called r and the first term of the sequence is 5, then the second term will equal $5r$ and the third term will equal $5r \cdot r = 5r^2$.

One way to solve this problem is to find the value of r that makes $5r^2$ equal to 245.

$$\frac{5r^2}{5} = \frac{245}{5}$$
$$r^2 = 49$$
$$r = \pm 7$$

There are two valid answers, but only one of them is among the four answer choices.

The correct choice is **(1)**.

Part II

25. To determine $g(-2)$, substitute -2 for x in the given equation.

$$g(-2) = -4(-2)^2 - 3(-2) + 2$$
$$= -4 \cdot 4 + 6 + 2$$
$$= -16 + 6 + 2$$
$$= -8$$

26. The commutative property of addition says that $a + b = b + a$. This can be applied to an expression involving negatives, like $-5a + 7 = 7 + (-5a) = 7 - 5a$.

The first step the student used was to rearrange the second and third terms of the right-hand side of the equation. So, the student used the commutative property of addition.

This is correct because using the commutative property of addition does not change the value of the right-hand side of the equation.

Most students would skip this as their first step and instead combine like terms to get $3a + 6 = 9 - 5a$. However, this commutative property step actually allows students to combine like terms.

27. Since there are no exponents greater than or equal to 2, this is a linear equation and the graph will be a line. Only two points are needed to graph a line. To be sure, though, make a chart with more values.

Pick any value you want for x, and substitute into the equation to solve for the corresponding y-value. For example, choose $x = 2$.

$$2y = -3(2) - 2$$
$$2y = -6 - 2$$
$$\frac{2y}{2} = \frac{-8}{2}$$
$$y = -4$$

x	y
0	−1
1	−2.5
2	−4
3	−5.5
4	−7

Plot the points on the graph, and connect them to draw the entire line.

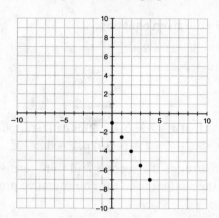 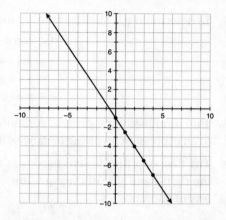

This could also have been done by first converting to slope-intercept form by dividing both sides of the equation by 2.

$$\frac{2y}{2} = \frac{-3x - 2}{2}$$

$$y = -\frac{3}{2}x - 1$$

The -1 is the y-intercept and the $-\frac{3}{2}$ is the slope. Graph the point $(0, -1)$. Then, using the slope of $-\frac{3}{2}$, move 2 units right and 3 units down to get the point $(2, -4)$. Join these two points to form the line.

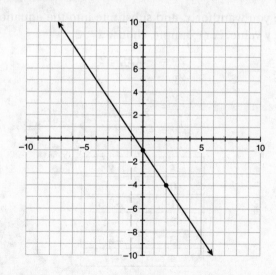

Since $(2, -4)$ is a point on the line, -4 is the value of k.

28. First multiply both sides of the equation by t.

$$at = v_f - v_i$$

Next, add v_i to both sides of the equation to isolate the v_f term.

$$at = v_f - v_i$$
$$+v_i = +v_i$$
$$at + v_i = v_f$$

29. Solve the inequality the same way you would solve it if there were an "$=$" sign instead of a "$<$." However, be sure to reverse the direction of the inequality sign when you multiply or divide both sides of the inequality by a negative number.

$$\frac{3}{5}x + \frac{1}{3} < \frac{4}{5}x - \frac{1}{3}$$

$$-\frac{1}{3} = -\frac{1}{3}$$

$$\frac{3}{5}x < \frac{4}{5}x - \frac{2}{3}$$

$$-\frac{4}{5}x = -\frac{4}{5}x$$

$$-\frac{1}{5}x < -\frac{2}{3}$$

$$-5\left(-\frac{1}{5}x\right) > -5\left(-\frac{2}{3}\right)$$

$$x > \frac{10}{3}$$

30. No. The product of two irrational numbers can be rational. For example, $\sqrt{2} \cdot \sqrt{2} = 2$. The number $\sqrt{2}$ is irrational, and the number 2 is rational.

31. Rearrange the terms to solve for x.

$$6x^2 - 42 = 0$$
$$+42 = +42$$
$$\frac{6x^2}{6} = \frac{42}{6}$$
$$x^2 = 7$$
$$\sqrt{x^2} = \pm\sqrt{7}$$
$$x = \pm\sqrt{7}$$

32. Make a table of values for the piecewise functions. For negative values, use the top equation. For values between 0 and 5, use the bottom equation.

x	h(x)
−2	$2(-2) - 3 = -7$
−1	$2(-1) - 3 = -5$
0	$0^2 - 4(0) - 5 = -5$
1	$1^2 - 4(1) - 5 = -8$
2	$2^2 - 4(2) - 5 = -9$
3	$3^2 - 4(3) - 5 = -8$
4	$4^2 - 4(4) - 5 = -5$
5	$5^2 - 4(5) - 5 = 0$

Plot these eight points with closed circles.

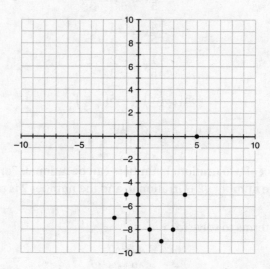

Before drawing the curves, evaluate the top equation at 0 to get $2(0) - 3 = -3$. Even though the point $(0, -3)$ is not part of the graph since 0 is part of the other "piece," this point is the boundary between the two pieces. Plot $(0, -3)$ with an open circle to represent this.

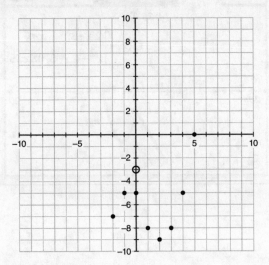

Now draw the curves.

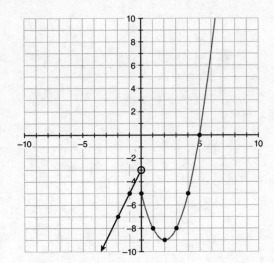

The TI-Nspire has the built-in capability for graphing piecewise functions, though it does not clearly show the open circle boundary point.

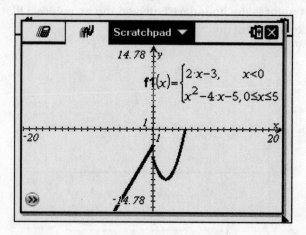

Part III

33. Rewrite both inequalities in slope-intercept form.

$$y \geq -2x + 8$$
$$y < 3x + 5$$

Graph the line $y = -2x + 8$ with a solid line because it contains the sign "$\geq$" and not "$>$." Test the ordered pair $(0, 0)$ to determine if it makes the inequality $y \geq -2x + 8$ true.

$$0 \geq -2(0) + 8$$
$$0 \geq 0 + 8$$
$$0 \geq 8$$

Since $(0, 0)$ does not make the first inequality true, shade the side of the line $y = -2x + 8$ that does not contain $(0, 0)$.

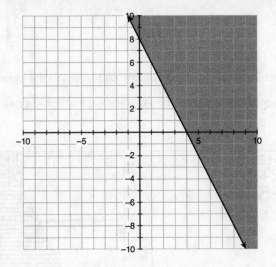

Graph the line $y = 3x + 5$ with a dotted line because it contains the sign "$<$" and not "$\leq$." Test the ordered pair $(0, 0)$ to determine if it makes the inequality $y = 3x + 5$ true.

$$0 < 3(0) + 5$$
$$0 < 0 + 5$$
$$0 < 5$$

Since (0, 0) does make the second inequality true, shade (with a different style of shading) the side of the line $y = 3x + 5$ that does contain (0, 0).

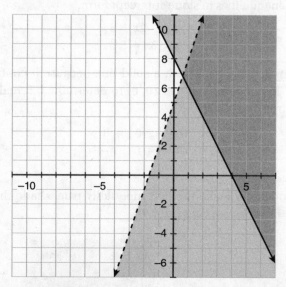

The entire double-shaded region is the solution set to the system of inequalities.

Systems of inequalities can also be graphed on a graphing calculator.

On the TI-84:

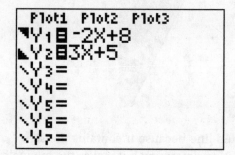
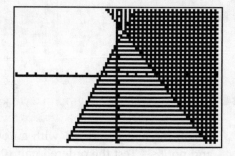

On the TI-Nspire:

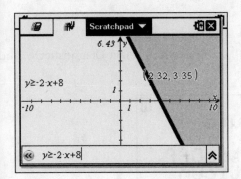
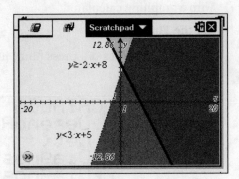

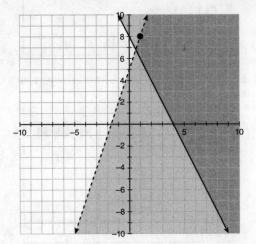

Since the point (1, 8) is on the dotted line, it is not part of the solution set.

34. The formula for exponential growth for a given growth rate is $A(t) = P(1 + r)^t$, where P is the original investment and r is the growth rate. For this problem, the function can be written as follows:

$$A(t) = 5000(1 + 0.012)^t = 5000(1.012)^t$$

Subtract the values when $t = 32$ and $t = 17$ to find the solution.

$$A(32) - A(17) = 5000(1.012)^{32} - 5000(1.012)^{17}$$
$$\approx 7324 - 6124$$
$$\approx 1200$$

To the *nearest dollar*, the investment will be worth $1200 more when Alexander turns 32 than when he turns 17.

35. A linear regression equation, or line of best fit, for a data set can be calculated with a graphing calculator.

For the TI-84:

To turn diagnostics on, press [2nd] [0], scroll down to select DiagnosticOn, and press [ENTER].

Press [STAT] [1]. Enter the *x*-values into L1 and the *y*-values into L2.

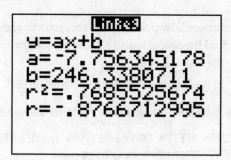

Press [STAT] [right arrow] [4] to get the LinReg function. Select Calculate.

For the TI-Nspire:

From the home screen, select the Add Lists & Spreadsheet icon. In the first row, label the A column "blocks" and the B column "dollars." Enter the blocks in cells A1 to A10. Enter the dollars in cells B1 to B10.

	A blocks	B dollars	C	D
◆				
1	0	293		
2	0	263		
3	1	244		
4	1	224		
5	3	185		

C1

Move the cursor to cell C1. Press [menu] [4] [1] [3]. Set X List to "blocks" and Y List to "dollars." Select OK.

	B dollars	C	D	E
◆				=LinRegM
1	293		Title	Linear Re..
2	263		RegEqn	m*x+b
3	244		m	-7.75635
4	224		b	246.338
5	185		r²	0.768553

E1 ="Linear Regression (mx+b)"

The line of best fit is $y = -7.76x + 246.34$.

The correlation coefficient (r) is a measure of how close the points are to the line of best fit. If the line of best fit has a positive slope, the correlation coefficient will be positive. If the line of best fit has a negative slope, the correlation coefficient will be negative. The closer the correlation coefficient is to 1 (or to -1), the closer to the line of best fit the points will lie.

For this data set, the correlation coefficient is -0.88. Because the correlation coefficient is negative, the line of best fit must have a negative slope.

36. Make a table of values for times 0 to 17, following the increases described in the problem.

Elapsed Time	Accumulated Snowfall
0	0
1	0.5
2	1
3	1.5
4	2
5	3
6	4
7	5
8	6
9	7
10	8
11	8
12	8
13	8
14	8.5
15	9
16	9.5
17	10

Plot those eighteen points, and connect the points to form the graph.

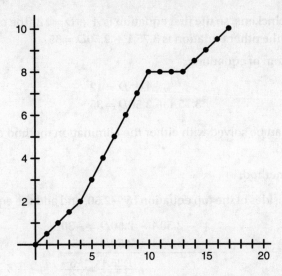

The average rate of snowfall is the total amount of snowfall divided by the total number of hours. For this problem, that is $\frac{10}{17} \approx 0.59$ inches per hour.

Part IV

37. There are 12 chickens, so the first equation is $A + D = 12$. The chickens cost 35 dollars, so the other equation is $3.75A + 2.50D = 35$.

Write the system of equations.

$$A + D = 12$$
$$3.75A + 2.50D = 35$$

This system can be solved with either the elimination method or with the substitution method.

Elimination method:

Multiply both sides of the top equation by -2.50, and add the equations together.

$$-2.50A - 2.50D = -30$$
$$+ \quad 3.75A + 2.50D = 35$$
$$\frac{1.25A}{1.25} = \frac{5}{1.25}$$
$$A = 4$$

Solve for D by plugging $A = 4$ into one of the original equations.

$$4 + D = 12$$
$$D = 8$$

Substitution method:

To use this method, one of the equations must be changed so that one of the variables is isolated. This can be done most easily with the first equation. Subtract D from both sides to make the equation $A = 12 - D$. Next, substitute into the second equation and solve for D.

$$3.75(12 - D) + 2.50D = 35$$
$$45 - 3.75D + 2.50D = 35$$
$$45 - 1.25D = 35$$
$$-45 = -45$$
$$\frac{-1.25D}{-1.25} = \frac{-10}{-1.25}$$
$$D = 8$$

Then solve for A by substituting $D = 8$ into the first equation.

$$A + 8 = 12$$
$$A = 4$$

Allysa bought 4 Americana chickens and 8 Delaware chickens.

With 4 Americana chickens and 8 Delaware chickens, Allysa will get $4 \cdot 2 + 8 \cdot 1 = 16$ eggs per day. After one week, she will have $16 \cdot 7 = 112$ eggs. Divide this by 12 to see how many dozens of eggs she will have in one week.

$$\frac{112}{12} \approx 9.3$$

Since she sells eggs only by the full dozen, Allysa can sell 9 dozen for a total of $9 \cdot \$2.50 = \$22.50$.

Topic	Question Numbers	Number of Points	Your Points	Your Percentage
1. Polynomials	3, 8, 12, 16	$2 + 2 + 2 + 2 = 8$		
2. Properties of Algebra	4, 26, 28	$2 + 2 + 2 = 6$		
3. Functions	2, 14, 21, 23, 25	$2 + 2 + 2 + 2 + 2 = 10$		
4. Creating and Interpreting Equations	1, 20	$2 + 2 = 4$		
5. Inequalities	29, 33	$2 + 4 = 6$		
6. Sequences and Series	19, 24	$2 + 2 = 4$		
7. Systems of Equations	22, 37	$2 + 6 = 8$		
8. Quadratic Equations and Factoring	5, 10, 11, 17, 31	$2 + 2 + 2 + 2 + 2 = 10$		
9. Regression	35	4		
10. Exponential Equations	7, 13, 34	$2 + 2 + 4 = 8$		
11. Graphing	18, 27, 32, 36	$2 + 2 + 2 + 4 = 10$		
12. Statistics	6, 15	$2 + 2 = 4$		
13. Number Properties	30	2		
14. Unit Conversions	9	2		

How to Convert Your Raw Score to Your Algebra I Regents Exam Score

The accompanying conversion chart must be used to determine your final score on the August 2019 Regents Exam in Algebra I. To find your final exam score, locate in the column labeled "Raw Score" the total number of points you scored out of a possible 86 points. Since partial credit is allowed in Parts II, III, and IV of the test, you may need to approximate the credit you would receive for a solution that is not completely correct. Then locate in the adjacent column to the right the scale score that corresponds to your raw score. The scale score is your final Algebra I Regents Exam score.

Regents Exam in Algebra I—August 2019
Chart for Converting Total Test Raw Scores
to Final Exam Scores (Scale Scores)

Raw Score	Scale Score	Performance Level	Raw Score	Scale Score	Performance Level	Raw Score	Scale Score	Performance Level
86	100	5	57	81	4	28	66	3
85	99	5	56	81	4	27	65	3
84	98	5	55	81	4	26	64	2
83	96	5	54	81	4	25	63	2
82	95	5	53	80	4	24	61	2
81	94	5	52	80	4	23	60	2
80	93	5	51	80	4	22	58	2
79	92	5	50	80	4	21	57	2
78	91	5	49	79	3	20	55	2
77	90	5	48	79	3	19	53	1
76	90	5	47	79	3	18	51	1
75	89	5	46	78	3	17	49	1
74	88	5	45	78	3	16	47	1
73	87	5	44	78	3	15	45	1
72	87	5	43	77	3	14	43	1
71	86	5	42	77	3	13	41	1
70	86	5	41	76	3	12	38	1
69	86	5	40	76	3	11	36	1
68	85	5	39	75	3	10	33	1
67	84	4	38	75	3	9	30	1
66	84	4	37	74	3	8	27	1
65	84	4	36	74	3	7	24	1
64	83	4	35	73	3	6	21	1
63	83	4	34	72	3	5	18	1
62	83	4	33	71	3	4	15	1
61	82	4	32	70	3	3	11	1
60	82	4	31	70	3	2	8	1
59	82	4	30	69	3	1	4	1
58	82	4	29	68	3	0	0	1

January 2020 Exam
Algebra I

High School Math Reference Sheet

Conversions

1 inch = 2.54 centimeters

1 meter = 39.37 inches

1 mile = 5280 feet

1 mile = 1760 yards

1 mile = 1.609 kilometers

1 kilometer = 0.62 mile

1 pound = 16 ounces

1 pound = 0.454 kilogram

1 kilogram = 2.2 pounds

1 ton = 2000 pounds

1 cup = 8 fluid ounces

1 pint = 2 cups

1 quart = 2 pints

1 gallon = 4 quarts

1 gallon = 3.785 liters

1 liter = 0.264 gallon

1 liter = 1000 cubic centimeters

Formulas

Triangle	$A = \frac{1}{2}bh$
Parallelogram	$A = bh$
Circle	$A = \pi r^2$
Circle	$C = \pi d$ or $C = 2\pi r$

Formulas (continued)

General Prisms	$V = Bh$
Cylinder	$V = \pi r^2 h$
Sphere	$V = \frac{4}{3}\pi r^3$
Cone	$V = \frac{1}{3}\pi r^2 h$
Pyramid	$V = \frac{1}{3}Bh$
Pythagorean Theorem	$a^2 + b^2 = c^2$
Quadratic Formula	$x = \dfrac{-b \pm \sqrt{b^2 - 4ac}}{2a}$
Arithmetic Sequence	$a_n = a_1 + (n - 1)d$
Geometric Sequence	$a_n = a_1 r^{n-1}$
Geometric Series	$S_n = \dfrac{a_1 - a_1 r^n}{1 - r}$ where $r \neq 1$
Radians	1 radian $= \frac{180}{\pi}$ degrees
Degrees	1 degree $= \frac{\pi}{180}$ radians
Exponential Growth/Decay	$A = A_0 e^{k(t-t_0)} + B_0$

PART I

Answer all 24 questions in this part. Each correct answer will receive 2 credits. No partial credit will be allowed. For each statement or question, write in the space provided the numeral preceding the word or expression that best completes the statement or answers the question.   [48 credits]

Please Note: Some topics will no longer be tested on future Regents Algebra I exams. These are noted with an asterisk next to the specific questions.

1. If $f(x) = 2(3^x) + 1$, what is the value of $f(2)$?

  (1)   13   (3)   37

  (2)   19   (4)   54   1   2

2. A high school sponsored a badminton tournament. After each round, one-half of the players were eliminated. If there were 64 players at the start of the tournament, which equation models the number of players left after 3 rounds?

  (1) $y = 64(1 - 0.5)^3$ (3) $y = 64(1 - 0.3)^{0.5}$

  (2) $y = 64(1 + 0.5)^3$ (4) $y = 64(1 + 0.3)^{0.5}$ 2   2

3. Given $7x + 2 \geq 58$, which number is *not* in the solution set?

  (1)   6   (3)   10

  (2)   8   (4)   12   3   2

4. Which table could represent a function?

x	f(x)
1	4
2	2
3	4
2	6

(1)

x	h(x)
2	6
0	4
1	6
2	2

(3)

x	g(x)
1	2
2	4
3	6
4	2

(2)

x	k(x)
2	2
3	2
4	6
3	6

(4)

4 _____

5. Which value of x makes $\frac{x-3}{4} + \frac{2}{3} = \frac{17}{12}$ true?

(1) 8   (3) 0

(2) 6   (4) 4

5 _____

6. Which expression is equivalent to $18x^2 - 50$?

(1) $2(3x+5)^2$ (3) $2(3x-5)(3x+5)$

(2) $2(3x-5)^2$ (4) $2(3x-25)(3x+25)$

6 _____

7. The functions $f(x) = x^2 - 6x + 9$ and $g(x) = f(x) + k$ are graphed below.

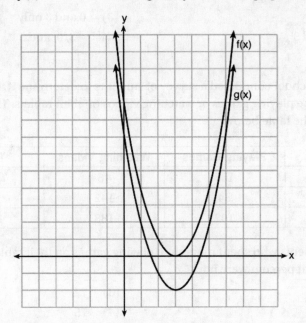

Which value of k would result in the graph of $g(x)$?

(1) 0

(2) 2

(3) −3

(4) −2

7 _2_

8. The shaded boxes in the figures below represent a sequence.

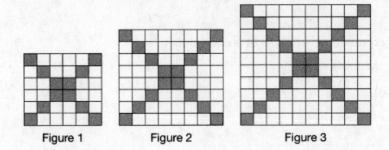

Figure 1   Figure 2   Figure 3

If figure 1 represents the first term and this pattern continues, how many shaded blocks will be in figure 35?

(1) 55

(2) 148

(3) 420

(4) 805

8 _2_

9. The zeros of the function $f(x) = x^3 - 9x^2$ are

(1)   9 only $x(x^2 - 9x)$

(2)   0 and 9 $x($

(3)   0 and 3 only

(4) $-3, 0,$ and 3

9 _____

10. A middle school conducted a survey of students to determine if they spent more of their time playing games or watching videos on their tablets. The results are shown in the table below.

	Playing Games	Watching Videos	Total
Boys	138	46	184
Girls	54	142	196
Total	192	188	380

Of the students who spent more time playing games on their tablets, approximately what percent were boys?

(1)   41

(2)   56

(3)   72

(4)   75

10 _____

11. Which statement best describes the solutions of a two-variable equation?

(1)   The ordered pairs must lie on the graphed equation.

(2)   The ordered pairs must lie near the graphed equation.

(3)   The ordered pairs must have $x = 0$ for one coordinate.

(4)   The ordered pairs must have $y = 0$ for one coordinate.

11 _____

12. The expression $x^2 - 10x + 24$ is equivalent to

(1) $(x + 12)(x - 2)$

(2) $(x - 12)(x + 2)$

(3) $(x + 6)(x + 4)$

(4) $(x - 6)(x - 4)$

12 _____

13. Which statement is true about the functions $f(x)$ and $g(x)$, given below?

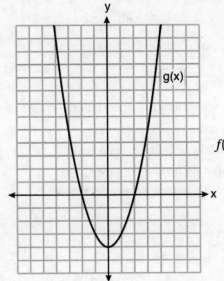

$$f(x) = -x^2 - 4x - 4$$

(1)   The minimum value of $g(x)$ is greater than the maximum value of $f(x)$. ✗

(2) $f(x)$ and $g(x)$ have the same y-intercept. ✗

(3) $f(x)$ and $g(x)$ have the same roots.

(4) $f(x) = g(x)$ when $x = -4$. ✗

13 _____

14. The equation $V(t) = 12{,}000(0.75)^t$ represents the value of a motorcycle t years after it was purchased. Which statement is true?

(1)   The motorcycle cost \$9000 when purchased.

(2)   The motorcycle cost \$12,000 when purchased.

(3)   The motorcycle's value is decreasing at a rate of 75% each year.

(4)   The motorcycle's value is decreasing at a rate of 0.25% each year.

14 _2___

15. The solutions to $(x + 4)^2 - 2 = 7$ are

(1) $-4 \pm \sqrt{5}$

(2) $4 \pm \sqrt{5}$ $(x+4)^2 = 9$

(3) -1 and -7

(4) 1 and 7

15 _2___

$x + 4 = \pm\sqrt{9}$

$x = -4 \pm 3$

16. Which expression is *not* equivalent to $-4x^3 + x^2 - 6x + 8$?

(1) $x^2(-4x + 1) - 2(3x - 4)$ $-4x^3 + 1x^2 - 6x + 8$

(2) $x(-4x^2 - x + 6) + 8$ $-4x^2 + x^2 + 6x + 8$

(3) $-4x^3 + (x - 2)(x - 4)$

(4) $-4(x^3 - 2) + x(x - 6)$

16 _____ 2

17. Which situation could be modeled as a linear equation?

(1)   The value of a car decreases by 10% every year.

(2)   The number of fish in a lake doubles every 5 years.

(3)   Two liters of water evaporate from a pool every day.

(4)   The amount of caffeine in a person's body decreases by $\frac{1}{3}$ every 2 hours.

17 _____ 3

18. The range of the function $f(x) = |x + 3| - 5$ is

(1) $[-5, \infty)$ (3) $[3, \infty)$

(2) $(-5, \infty)$ (4) $(3, \infty)$

18 _____ 2

19. A laboratory technician used the function $t(m) = 2(3)^{2m+1}$ to model her research. Consider the following expressions:

          I. $6(3)^{2m}$ II. $6(6)^{2m}$ III. $6(9)^{m}$

The function $t(m)$ is equivalent to

(1)   I only   (3)   I and III

(2)   II only   (4)   II and III

19 _____ 3

***20.** Which system of equations has the same solutions as the system below?

$$3x - y = 7$$
$$2x + 3y = 12$$

(1) $6x - 2y = 14$ (3) $-9x - 3y = -21$
      $-6x + 9y = 36$ $2x + 3y = 12$

(2) $18x - 6y = 42$ (4) $3x - y = 7$
      $4x + 6y = 24$ $x + y = 2$

20 _____ 2

*This topic will no longer be tested on future Regents Algebra I exams.

---

21. A population of paramecia, P, can be modeled using the exponential function $P(t) = 3(2)^t$, where t is the number of days since the population was first observed. Which domain is most appropriate to use to determine the population over the course of the first two weeks?

(1) $t \geq 0$ (3) $0 \leq t \leq 2$

(2) $t \leq 2$ (4) $0 \leq t \leq 14$ 21 _____

22. Given the following data set:

$$65, 70, 70, 70, 70, 80, 80, 80, 85, 90, 90, 95, 95, 95, 100$$

Which representations are correct for this data set?

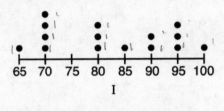

I

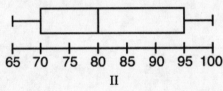

II

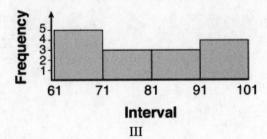

III

(1)   I and II only   (3)   II and III only

(2)   I and III only   (4)   I, II, and III   22 _____

*23. A recursively defined sequence is shown below.

$$a_1 = 5$$
$$a_{n+1} = 2a_n - 7$$

The value of a_4 is

(1) -9　　(3) 8

(2) -1　　(4) 15　　23 ______

24. Which polynomial has a leading coefficient of 4 and a degree of 3?

(1) $3x^4 - 2x^2 + 4x - 7$　　(3) $4x^4 - 3x^3 + 2x^2$

(2) $4 + x - 4x^2 + 5x^3$　　(4) $2x + x^2 + 4x^3$　　24 ______

*This topic will no longer be tested on future Regents Algebra I exams.

PART II

Answer all 8 questions in this part. Each correct answer will receive 2 credits. Clearly indicate the necessary steps, including appropriate formula substitutions, diagrams, graphs, charts, etc. For all questions in this part, a correct numerical answer with no work shown will receive only 1 credit. [16 credits]

25. Graph $f(x) = -\sqrt{x} + 1$ on the set of axes below.

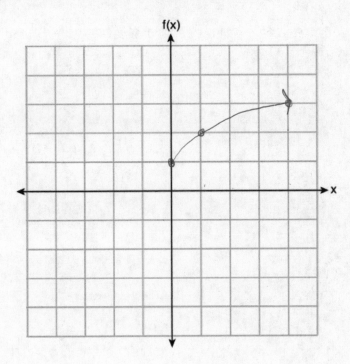

26. Maria orders T-shirts for her volleyball camp. Adult-sized T-shirts cost $6.25 each and youth-sized T-shirts cost $4.50 each. Maria has $550 to purchase both adult-sized and youth-sized T-shirts. If she purchases 45 youth-sized T-shirts, determine algebraically the maximum number of adult-sized T-shirts she can purchase.

$$6.25x + 45(4.50) \leq 550$$

$$6.25x + 202.50 \leq 550$$

$$6.25x \leq 550 - 202.50$$

$$6.25x \leq 347.50$$

$$x \leq 55.6$$

$$\boxed{x \leq 55}$$

2

27. A news report suggested that an adult should drink a minimum of 4 pints of water per day. Based on this report, determine the *minimum* amount of water an adult should drink, in fluid ounces, per week.

≤ 4

28. Express $(3x - 4)(x + 7) - \frac{1}{4}x^2$ as a trinomial in standard form.

$$3x^2 + 21x - 4x - 28 - \frac{1}{4}x^2$$

$$2.75x^2 + 17x - 28$$

29. John was given the equation $4(2a + 3) = -3(a - 1) + 31 - 11a$ to solve. Some of the steps and their reasons have already been completed. State a property of numbers for each missing reason.

$$4(2a + 3) = -3(a - 1) + 31 - 11a \qquad \text{Given}$$

$$8a + 12 = -3a + 3 + 31 - 11a \qquad \underline{\text{distribute}}$$

$$8a + 12 = 34 - 14a \qquad \text{Combining like terms}$$

$$22a + 12 = 34 \qquad \underline{\text{commutative add.}}$$

30. State whether the product of $\sqrt{3}$ and $\sqrt{9}$ is rational or irrational. Explain your answer.

$$\sqrt{3} \cdot \sqrt{9} = 5.196\ldots$$

irrational

non repeating non terminating decimal

2

31. Use the method of completing the square to determine the exact values of x for the equation $x^2 - 8x + 6 = 0$.

$$x^2 - 8x = -6$$

$$x^2 - 8x + 16 = -6 + 16$$

$$x^2 - 8x + 16 = 10$$

$$(x - 4)^2 = 10$$

$$x - 4 = \pm\sqrt{10}$$

$$x = 4 \pm \sqrt{10}$$

2

32. A formula for determining the finite sum, S, of an arithmetic sequence of numbers is $S = \frac{n}{2}(a + b)$, where n is the number of terms, a is the first term, and b is the last term. Express b in terms of a, S, and n.

$$\frac{S - a}{\frac{n}{2}} = b$$

PART III

Answer all 4 questions in this part. Each correct answer will receive 4 credits. Clearly indicate the necessary steps, including appropriate formula substitutions, diagrams, graphs, charts, etc. For all questions in this part, a correct numerical answer with no work shown will receive only 1 credit. [16 credits]

33. Michael threw a ball into the air from the top of a building. The height of the ball, in feet, is modeled by the equation $h = -16t^2 + 64t + 60$, where t is the elapsed time, in seconds. Graph this equation on the set of axes below.

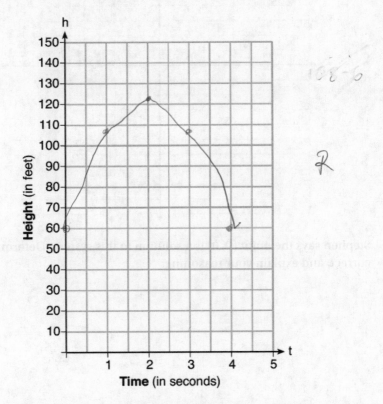

Determine the average rate of change, in feet per second, from when Michael released the ball to when the ball reached its maximum height.

$$\frac{48}{2} \qquad \frac{108-60}{2-0}$$

$$\boxed{24}$$

34. Graph the system of inequalities:

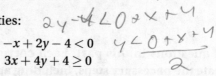

$$-x + 2y - 4 < 0$$
$$3x + 4y + 4 \geq 0$$

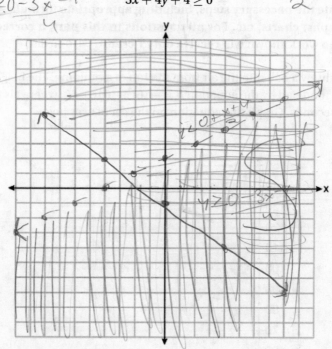

Stephen says the point (0, 0) is a solution to this system. Determine if he is correct, and explain your reasoning.

yes. both areas are shaded
at that cordinate

35. The following table represents a sample of sale prices, in thousands of dollars, and number of new homes available at that price in 2017.

Sale Price, p (in thousands of dollars)	160	180	200	220	240	260	280
Number of New Homes Available $f(p)$	126	103	82	75	82	40	20

State the linear regression function, $f(p)$, that estimates the number of new homes available at a specific sale price, p. Round all values to the *nearest hundredth*.

$$y = -.74x + 249.86$$

State the correlation coefficient of the data to the *nearest hundredth*. Explain what this means in the context of the problem.

$$-.95$$

Strong negative correlation

36. The length of a rectangular sign is 6 inches more than half its width. The area of this sign is 432 square inches. Write an equation in one variable that could be used to find the number of inches in the dimensions of this sign.

$$6 + \frac{1}{2}w = L$$

$$w \cdot L = 432$$

Solve this equation algebraically to determine the dimensions of this sign, in inches.

$$w \cdot \left(6 + \frac{1}{2}w\right) = L$$

$$6w + \frac{1}{2}w^2 = 432$$

$$6w + \frac{1}{2}w^2 - 432 = 0$$

$$\frac{-\frac{1}{2} \pm \sqrt{\frac{1}{2}^2 - 4(6)(-432)}}{2(6)}$$

$$\frac{-.5 \pm \sqrt{.25 + 10368}}{12}$$

$$\frac{-.5 \pm \sqrt{10368.25}}{12}$$

PART IV

Answer the question in this part. A correct answer will receive 6 credits. Clearly indicate the necessary steps, including appropriate formula substitutions, diagrams, graphs, charts, etc. A correct numerical answer with no work shown will receive only 1 credit.   [6 credits]

37. Two families went to Rollercoaster World. The Brown family paid $170 for 3 children and 2 adults. The Peckham family paid $360 for 4 children and 6 adults.

    If x is the price of a child's ticket in dollars and y is the price of an adult's ticket in dollars, write a system of equations that models this situation.

$$3x + 2y = 470$$
$$4x + 6y = 360$$

Graph your system of equations on the set of axes below.

$$y = \frac{170 - 3x}{2}$$

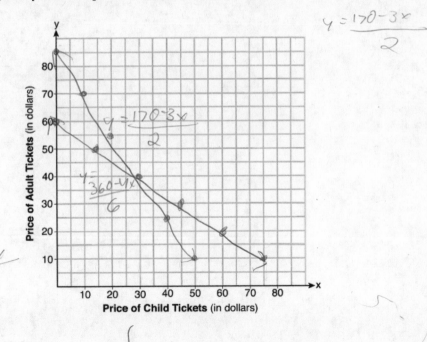

$$\frac{360 - 4x}{6}$$

6

State the coordinates of the point of intersection.

(30, 40)

Explain what each coordinate of the point of intersection means in the context of the problem.

when a child ticket is 30 $,
an adult ticket is 40 $

Answers January 2020
Algebra I

Answer Key

PART I

1. (2)	5. (2)	9. (2)	13. (2)	17. (3)	21. (4)
2. (1)	6. (3)	10. (3)	14. (2)	18. (1)	22. (4)
3. (1)	7. (4)	11. (1)	15. (3)	19. (3)	23. (1)
4. (2)	8. (2)	12. (4)	16. (2)	20. (2)	24. (4)

PART II

25.

26. 55 T-shirts

27. 448 fluid ounces

28. $\frac{11}{4}x^2 + 17x - 28$

29. The distributive property and the addition property of equality

30. Irrational

31. $4 \pm \sqrt{10}$

32. $\frac{2S}{n} - a = b$

PART III

33. 32 feet per second

34. Stephen is correct that $(0, 0)$ is a solution to this system.

35. $f(p) = -0.79p + 249.86$; $r = -0.95$

36. $432 = \left(\frac{1}{2}w + 6\right)(w)$; $w = 24$ inches; $l = 18$ inches

PART IV

37. $3x + 2y = 170$;
$4x + 6y = 360$;
$(30, 40)$

In **Parts II–IV**, you are required to show how you arrived at your answers. For sample methods of solutions, see the *Answer Explanations* section.

Answer Explanations

Part I

Please Note: Some topics will no longer be tested on future Regents Algebra I exams. These are noted with an asterisk next to the specific questions.

1. To evaluate the function $f(x) = 2(3^x) + 1$ at $x = 2$, substitute 2 for the value of x in the equation.

$$f(x) = 2(3^x) + 1$$
$$f(2) = 2(3^{(2)}) + 1$$
$$f(2) = 2(9) + 1$$
$$f(2) = 18 + 1$$
$$f(2) = 19$$

The correct choice is **(2)**.

2. According to the problem, the number of players decreases by a constant percent rate per unit interval. Therefore, this problem can be modeled with a population decay equation of the form $y = C(1 - r)^t$, where y is the population at time t, C is the initial population, and r is the decay rate. In this case, $C = 64$ because there were 64 players at the start of the tournament, $r = 0.5$ because the population decreases by half each round, and $t = 3$ because this went on for 3 rounds. This makes the equation $y = 64(1 - 0.5)^3$.

The correct choice is **(1)**.

3. Test each of the possibilities in the inequality.

Testing choice (2):

$$7(8) + 2 = 58 \geq 58$$

Testing choice (3):

$$7(10) + 2 = 72 \geq 58$$

Testing choice (4):

$$7(12) + 2 = 86 \geq 58$$

The inequality does hold for choices (2), (3), and (4). It does not, however, hold for choice (1).

$$7(6) + 2 = 44 \ngeq 58$$

The correct choice is **(1)**.

4. A function is a relation for which $f(x)$ is defined uniquely for each value of x.

Choice (1) is not a function because $f(2) = 2$ but also $f(2) = 6$.

Choice (3) is not a function because $h(2) = 6$ but also $h(2) = 2$.

Choice (4) is not a function because $k(3) = 2$ but also $k(3) = 6$.

Choice (2), on the other hand, is a function because each value of x maps onto exactly one value of $g(x)$.

The correct choice is **(2)**.

5. Test each of the possibilities in the equality.

Testing choice (1):

$$\frac{(8) - 3}{4} + \frac{2}{3} \overset{?}{=} \frac{17}{12}$$

$$\frac{5}{4} + \frac{2}{3} \overset{?}{=} \frac{17}{12}$$

$$\frac{15}{12} + \frac{8}{12} \overset{?}{=} \frac{17}{12}$$

$$\frac{23}{12} \neq \frac{17}{12}$$

Testing choice (2):

$$\frac{(6) - 3}{4} + \frac{2}{3} \overset{?}{=} \frac{17}{12}$$

$$\frac{3}{4} + \frac{2}{3} \overset{?}{=} \frac{17}{12}$$

$$\frac{9}{12} + \frac{8}{12} \overset{?}{=} \frac{17}{12}$$

$$\frac{17}{12} \overset{\checkmark}{=} \frac{17}{12}$$

Testing choice (3):

$$\frac{(0)-3}{4}+\frac{2}{3} \stackrel{?}{=} \frac{17}{12}$$

$$\frac{-3}{4}+\frac{2}{3} \stackrel{?}{=} \frac{17}{12}$$

$$\frac{-9}{12}+\frac{8}{12} \stackrel{?}{=} \frac{17}{12}$$

$$\frac{-1}{12} \neq \frac{17}{12}$$

Testing choice (4):

$$\frac{(4)-3}{4}+\frac{2}{3} \stackrel{?}{=} \frac{17}{12}$$

$$\frac{1}{4}+\frac{2}{3} \stackrel{?}{=} \frac{17}{12}$$

$$\frac{3}{12}+\frac{8}{12} \stackrel{?}{=} \frac{17}{12}$$

$$\frac{11}{12} \neq \frac{17}{12}$$

The correct choice is **(2)**.

6. Start by noticing that both 18 and 50 are divisible by 2. This means that 2 can be factored out of the expression.

$$18x^2 - 50 = 2(9x^2 - 25)$$

Next, notice that $9x^2 - 25$ is the difference of perfect squares because $9x^2 = (3x)^2$ and $25 = 5^2$. With the difference of perfect squares, $(a^2 - b^2) = (a - b)(a + b)$. Therefore:

$$2(9x^2 - 25) = 2(3x - 5)(3x + 5)$$

The correct choice is **(3)**.

7. The addition of a constant k to a function translates the graph up or down the y-axis by k units. The graph shows that $g(x)$ is $f(x)$ translated by -2 units.

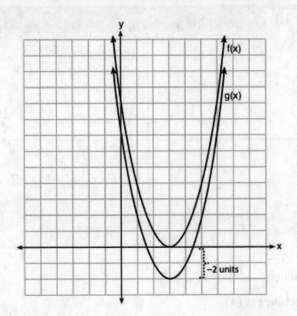

Since the function $f(x)$ is explicitly given, check your work by graphing $g(x) = f(x) - 2 = x^2 - 6x + 9 - 2$ on the graphing calculator.

For the TI-84:

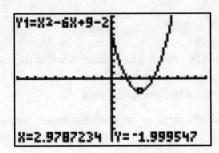

The graph matches $g(x)$.

For the TI-Nspire:

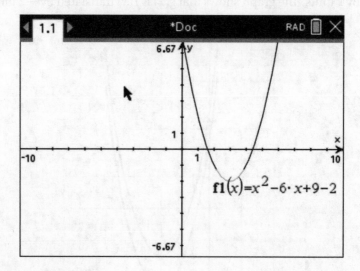

The graph matches $g(x)$.

The correct choice is **(4)**.

8. Count the number of shaded blocks in each of the three figures. Let a_n equal the number of shaded blocks in Figure n.

$$a_1 = 12$$
$$a_2 = 16$$
$$a_3 = 20$$

An arithmetic sequence is one where each term is the same amount more than the previous term. For this sequence, the first term of this sequence is 12 and each term is 4 more than the previous term.

The formula for the nth term of an arithmetic sequence is $a_n = a_1 + (n-1)d$, where a_1 is the first term, n is the term number, and d is the common difference.

To find the 35th term of this sequence:

$$
\begin{aligned}
a_{35} &= 12 + (35 - 1)(4) \\
&= 12 + (34)(4) \\
&= 12 + 136 \\
&= 148
\end{aligned}
$$

The correct choice is **(2)**.

9. The zeros of a function $f(x)$ are the values of x for which $f(x) = 0$.

$$f(x) = 0$$
$$x^3 - 9x^2 = 0$$

Factor x out of the equation twice.

$$(x)(x)(x - 9) = 0$$

The equation is satisfied whenever one of the factors equals 0. For the first two factors, this is trivial.

$$x = 0$$

For the third factor:

$$x - 9 = 0$$
$$x = 9$$

The zeros of the function are 0 and 9.

The correct choice is **(2)**.

10. To calculate the percentage, construct a fraction. The denominator of the fraction is the total number of students who spent more time playing games on their tablets. According to the table, this is 192. The numerator of the fraction is the number of boys who spent more time playing games on their tablets. According to the table, this is 138.

The fraction is $\frac{138}{192} = 0.71875$. Convert this to a percentage by multiplying by 100 and rounding.

$$0.71875 \times 100 = 71.875 \approx 72$$

The correct choice is **(3)**.

11. The solutions of a two-variable equation are the set of ordered pairs that satisfy the equation. When graphed, these ordered pairs form the graph of the equation.

The ordered pair where $x = 0$ is the y-intercept of the equation. The ordered pair where $y = 0$ is called the zero, the root, or the x-intercept.

The correct choice is **(1)**.

12. Factor the expression using the reverse FOIL method by finding two numbers whose sum is -10 and whose product is 24. Notice that the answer choices provide possible factors. In particular, notice that $(-6) + (-4) = -10$ and $(-6)(-4) = 24$. This suggests that $(x - 6)$ and $(x - 4)$ are factors.

Check your work by multiplying $(x - 6)$ and $(x - 4)$ using FOIL and simplifying.

$$(x - 6)(x - 4) = x^2 - 4x - 6x + 24$$
$$= x^2 - 10x + 24$$

The correct choice is **(4)**.

13. Graph $f(x)$ on the graphing calculator.

For the TI-84:

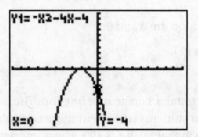

For the TI-Nspire:

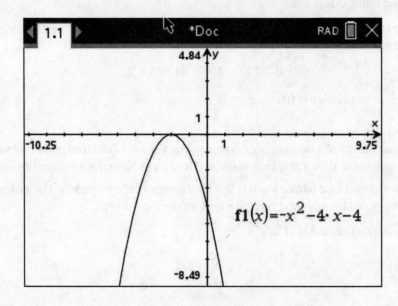

Compare the graph of $f(x)$ to that of $g(x)$. The minimum value of $g(x)$ is -4, while $f(x)$ decreases unbounded as x approaches positive and negative infinity. Choice (1) is not correct.

The y-intercept of $g(x)$ is -4, and the y-intercept of $f(x)$ is also -4. Choice (2) is correct.

The roots of $g(x)$ are where the graph intersects the x-axis. This occurs at $x = 2$ and $x = -2$. For $f(x)$, the graph intersects the x-axis at $x = -2$ but not at $x = 2$. Therefore, choice (3) is not correct.

$g(-4) = 12$ while $f(-4) = -4$. Therefore, choice (4) is not correct.

The correct choice is **(2)**.

14. Interpret the equation using the compound interest formula, $y = a(1 + r)^x$, where y is the money at any given time, a is the initial deposit, r is the rate of interest, and x is the number of years. In the equation $V(t) = 12{,}000(0.75)^t$, $a = 12{,}000$ and $r = -0.25$.

Thus, the motorcycle cost \$12,000 when purchased. Note that choice (4) is not correct because when 0.25% is rewritten as a decimal, it is equal to 0.0025.

The correct choice is **(2)**.

15. Find the solutions by isolating x.

$$(x + 4)^2 - 2 = 7$$
$$(x + 4)^2 = 9$$
$$\sqrt{(x + 4)^2} = \sqrt{9}$$

Notice that $(3)^2 = 9$ and $(-3)^2 = 9$. Examine the solution in the first case.

$$x + 4 = 3$$
$$x = -1$$

Now examine the solution in the second case.

$$x + 4 = -3$$
$$x = -7$$

The correct choice is **(3)**.

16. Test each choice by simplifying.

Testing choice (1):

$$x^2(-4x + 1) - 2(3x - 4) = -4x^3 + x^2 - 6x + 8$$

Testing choice (2):

$$x(-4x^2 - x + 6) + 8 = -4x^3 - x^2 + 6x + 8$$

Testing choice (3):

$$-4x^3 + (x - 2)(x - 4) = -4x^3 + x^2 - 4x - 2x + 8 = -4x^3 + x^2 - 6x + 8$$

Testing choice (4):

$$-4(x^3 - 2) + x(x - 6) = -4x^3 + 8 + x^2 - 6x = -4x^3 + x^2 - 6x + 8$$

The correct choice is **(2)**.

17. A linear equation models scenarios where a dependent variable increases or decreases by the same amount for each one unit of change in the independent variable.

For choice (1), suppose a car is worth $10,000 at the beginning. At the end of the first year, its price will decrease by 10%, or $1000, and it will be worth $9000. At the end of the second year, its price will decrease by 10%, or $900, and it will be worth $8100. Choice (1) is not correct because the amount of decrease is different between year 1 and year 2.

For choice (2), suppose a lake has 2 fish at the beginning. After the first 5 years, the population will double and increase by 2, meaning that there will be 4 fish in the lake. At the end of the next 5-year period, the population will double and increase by 4, meaning that there will be 8 fish in the lake. Choice (2) is not correct because the amount of increase is different between the two 5-year periods.

For choice (3), suppose a pool has 100 L of water at the beginning. At the end of the first day, 2 L will have evaporated, leaving 98 L in the pool. At the end of the second day, 2 L will have evaporated, leaving 96 L in the pool. Choice (3) is correct because the same amount of water evaporates every day.

For choice (4), suppose a person starts with 81 mg of caffeine in their body. At the end of 2 hours, $\frac{1}{3}$ of 81 mg, or 27 mg, will have left the body, leaving 54 mg remaining. At the end of the next 2 hours, $\frac{1}{3}$ of 54 mg, or 18 mg, will have left the body, leaving 36 mg remaining. Choice (4) is not correct because the amount of decrease between the first 2-hour period and the second 2-hour period is not the same.

The correct choice is **(3)**.

18. The range is the set of y-coordinates of all the points on the graph of a function. Find the range by graphing.

On the TI-84:

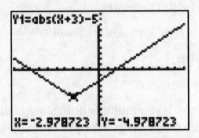

On the TI-Nspire:

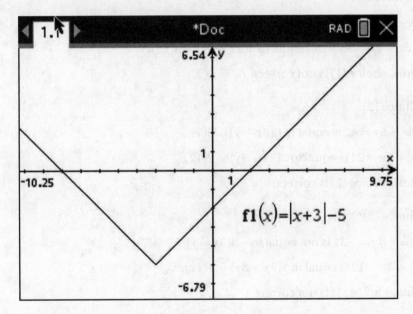

The lowest y-coordinate of the graph is -5. There is no highest y-coordinate since the graph increases unbounded as x approaches both positive and negative infinity. Since -5 is included in the range, use the square bracket to show inclusion of that point.

The correct choice is (1).

19. Use the properties of exponents to simplify $t(m) = 2(3)^{2m+1}$. Recall that $a^{x+y} = a^x a^y$. Therefore:

$$2(3)^{2m+1} = 2(3)^{2m}(3)^1 = 6(3)^{2m}$$

Also recall that $a^{xy} = (a^x)^y$. Therefore:

$$6(3)^{2m} = 6(3^2)^m = 6(9)^m$$

The correct choice is **(3)**.

*20. In a system of equations, if you multiply both sides of one (or both) of the equations by a constant, the new system will have the same solutions as the original system. Look at each of the four choices.

Choice (1):

$6x - 2y = 14$ is equal to $2(3x - y) = 2(7)$.

$-6x + 9y = 36$ is not equal to $3(2x + 3y) = 3(12)$.

Thus, choice (1) is not correct.

Choice (2):

$18x - 6y = 42$ is equal to $6(3x - y) = 6(7)$.

$4x + 6y = 24$ is equal to $2(2x + 3y) = 2(12)$.

Thus, choice (2) is correct.

Choice (3):

$-9x - 3y = -21$ is not equal to $-3(3x - y) = -3(7)$.

$2x + 3y = 12$ is equal to $1(2x + 3y) = 1(12)$.

Thus, choice (3) is not correct.

Choice (4):

$3x - y = 7$ is equal to $1(3x - y) = 1(7)$.

$x + y = 2$ is not equal to $\frac{1}{2}(2x + 3y) = \frac{1}{2}(12)$.

Thus, choice (4) is not correct.

The correct choice is **(2)**.

*This topic will no longer be tested on future Regents Algebra I exams.

21. The domain is the set of t-values for which the model makes sense. The question tells you that t is the number of days since the population was first observed, and you are interested in the population over the course of two weeks. Therefore, you are interested in the model between 0 and 14 days. This is represented by the range $0 \leq t \leq 14$.

The correct choice is **(4)**.

22. The question asks you to represent the data set as a dot plot, a box plot, and as a histogram.

To make a dot plot, draw a number line and plot the frequency of each value as a dot above the number on the number line. According to the data set, the frequency of each value is the following:

Value	Frequency
65	1
70	4
75	0
80	3
85	1
90	2
95	3
100	1

A dot plot of the table is exactly that shown in I.

To make a box plot, find the minimum, first quartile, median, third quartile, and maximum.

The minimum of the data set is 65. The maximum of the data set is 100.

The first quartile is the value for which a quarter of the data is less than that value. The median is the value for which half of the data is less than that value. The third quartile is the value for which three-quarters of the data is less than that value. To determine these three values, write out the values in the data set from lowest to highest. Then circle the 4th, 8th, and 12th values in the data set.

First quartile   Median   Third quartile

65, 70, 70, (70,) 70, 80, 80, (80,) 85, 90, 90, (95,) 95, 95, 100

Notice that there are 3 data points below the first quartile, 3 between the first quartile and the median, 3 between the median and the third quartile, and 3 above the third quartile.

The values 65, 70, 80, 95, and 100 are represented in II.

To make a histogram, bin the data into ranges.

Value	Frequency
61–70	5
71–80	3
81–90	3
91–100	4

A histogram of the table is exactly that shown in III.

The correct choice is **(4)**.

*23. To get a_4, calculate each term of the recursive function. We are told:

$$a_1 = 5$$

Using the recursive function, calculate a_2.

$$a_2 = 2(a_1) - 7$$
$$= 2(5) - 7$$
$$= 3$$

---

*This topic will no longer be tested on future Regents Algebra I exams.

Using the recursive function, calculate a_3.

$$a_3 = 2(a_2) - 7$$
$$= 2(3) - 7$$
$$= -1$$

Using the recursive function, calculate a_4.

$$a_4 = 2(a_3) - 7$$
$$= 2(-1) - 7$$
$$= -9$$

The correct choice is **(1)**.

24. The leading coefficient of a polynomial is the coefficient of the variable x raised to the highest power. The degree of a polynomial is the highest power of x appearing in the polynomial. To answer this question, identify the term with the highest power of x in each polynomial.

For choice (1), the term with the highest power of x in $3x^4 - 2x^2 + 4x - 7$ is $3x^4$. The leading coefficient is 3, and the degree is 4. Thus, choice (1) is not correct.

For choice (2), the term with the highest power of x in $4 + x - 4x^2 + 5x^3$ is $5x^3$. The leading coefficient is 5, and the degree is 3. Thus, choice (2) is not correct.

For choice (3), the term with the highest power of x in $4x^4 - 3x^3 + 2x^2$ is $4x^4$. The leading coefficient is 4, and the degree is 4. Thus, choice (3) is not correct.

For choice (4), the term with the highest power of x in $2x + x^2 + 4x^3$ is $4x^3$. The leading coefficient is 4, and the degree is 3.

The correct choice is **(4)**.

Part II

25. Make a table of values for $f(x)$. Because this function has a square root, choose perfect squares to help with the arithmetic.

x	f(x)
−2	$-\sqrt{-2} + 1 =$ undefined
−1	$-\sqrt{-1} + 1 =$ undefined
0	$-\sqrt{0} + 1 = 1$
1	$-\sqrt{1} + 1 = 0$
4	$-\sqrt{4} + 1 = -1$
9	$-\sqrt{9} + 1 = -2$

Plot these points with closed circles.

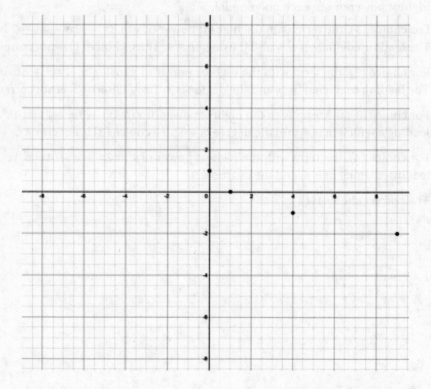

Connect the points to draw the curve.

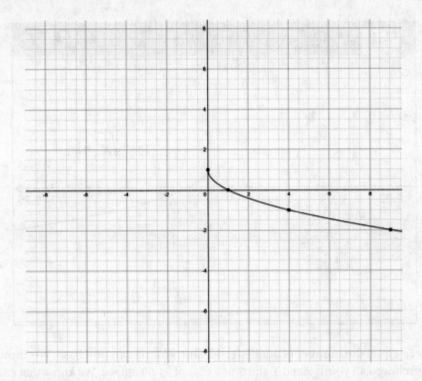

The graphs can be plotted on a graphing calculator to check your work.

For the TI-84:

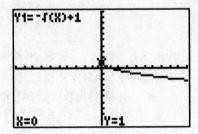

For the TI-Nspire:

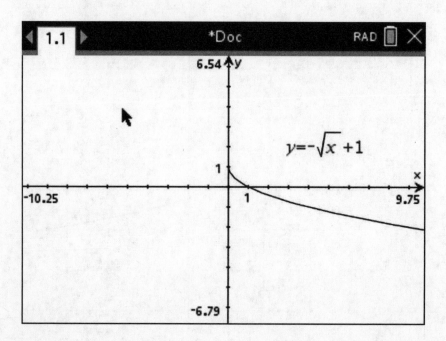

26. Let x equal the number of adult-sized T-shirts Maria can purchase. We know she purchased 45 youth-sized T-shirts at a cost of \$4.50 apiece. We know that each adult-sized T-shirt costs \$6.25. We also know that the sum of all the T-shirts purchased is less than or equal to \$550. Therefore, an equation that can model this situation is the following:

$$(45)(\$4.50) + (x)(\$6.25) \leq \$550$$

Solve for x.

$$(45)(\$4.50) + (x)(\$6.25) \leq \$550$$
$$\$202.50 + (x)(\$6.25) \leq \$550$$
$$(x)(\$6.25) \leq \$347.50$$
$$x \leq 55.6$$

Since T-shirts can be sold only as whole shirts, look for the largest integer that satisfies the inequality. The maximum number of adult-sized T-shirts that Maria can buy is 55 T-shirts.

27. Use the reference sheet to determine the number of fluid ounces in a pint. According to the reference sheet, 1 pint = 2 cups and 1 cup = 8 fluid ounces.

In addition, determine the number of days in a week: 1 week = 7 days.

Use these ratios to convert 4 pints per day to fluid ounces per week.

$$\left(\frac{4 \text{ pints}}{\text{day}}\right) \times \left(\frac{2 \text{ cups}}{1 \text{ pint}}\right) \times \left(\frac{8 \text{ fluid ounces}}{1 \text{ cup}}\right) \times \left(\frac{7 \text{ days}}{1 \text{ week}}\right)$$

Notice that the units pints, cups, and days cancel out. Now simplify:

$$\left(\frac{4 \text{ pints}}{\text{day}}\right) \times \left(\frac{2 \text{ cups}}{1 \text{ pint}}\right) \times \left(\frac{8 \text{ fluid ounces}}{1 \text{ cup}}\right) \times \left(\frac{7 \text{ days}}{1 \text{ week}}\right) = \frac{448 \text{ fluid ounces}}{\text{week}}$$

An adult should drink a minimum of 448 fluid ounces of water per week.

28. Simplify the expression using the distributive and commutative properties.

$$(3x - 4)(x + 7) - \tfrac{1}{4}x^2 = 3x^2 + 21x - 4x - 28 - \tfrac{1}{4}x^2$$

$$= \left(3 - \tfrac{1}{4}\right)x^2 + (21 - 4)x - 28$$

$$= \tfrac{11}{4}x^2 + 17x - 28$$

29. The distributive property states that $a(b + c) = ab + ac$. In step 2 of the solution, John simplified $4(2a + 3) = -3(a - 1) + 31 - 11a$ to $8a + 12 = -3a + 3 + 31 - 11a$. This is an application of the distributive property.

The addition property of equality states that adding equal quantities to both sides of an equation produces another equivalent equation. In step 4 of the solution, John added $14a$ to both sides of the equation from step 3.

$$8a + 12 + 14a = 34 - 14a + 14a$$
$$22a + 12 = 34$$

30. The quantity $\sqrt{3} \approx 1.7320508$ is an irrational number because the decimal does not terminate and does not repeat. On the other hand, $\sqrt{9} = 3$ or -3 are both rational numbers because their decimals terminate.

The product of an irrational number and a rational number is always irrational. Therefore, the product of $\sqrt{3}$ and $\sqrt{9}$ is irrational.

31. To complete the square, first eliminate the constant from the left side of the equation by subtracting 6 from both sides.

$$x^2 - 8x + 6 = 0$$
$$x^2 - 8x = -6$$

Then, divide the coefficient of x by 2, square that answer, and add the number to both sides of the equation. The coefficient of x is -8.

$$\left(-\frac{8}{2}\right)^2 = (-4)^2 = 16$$

Add 16 to both sides of the equation.

$$x^2 - 8x + 16 = -6 + 16$$
$$x^2 - 8x + 16 = 10$$

Factor the left side of the equation as a perfect square.

$$x^2 - 8x + 16 = 10$$
$$(x - 4)^2 = 10$$

Take the square root of both sides of the equation.

$$(x - 4)^2 = 10$$
$$\sqrt{(x - 4)^2} = \pm\sqrt{10}$$
$$x - 4 = \pm\sqrt{10}$$
$$x = 4 \pm \sqrt{10}$$

32. To express b in terms of a, S, and n, isolate b.

$$S = \frac{n}{2}(a + b)$$

$$\left(\frac{2}{n}\right)S = \left(\frac{2}{n}\right)\frac{n}{2}(a + b)$$

$$\left(\frac{2}{n}\right)S = a + b$$

$$\frac{2S}{n} = a + b$$

$$\frac{2S}{n} - a = b$$

Part III

33. Make a table of values for $h(t)$.

t	h(t)
0	$-16(0)^2 + 64(0) + 60 = 60$
1	$-16(1)^2 + 64(1) + 60 = 108$
2	$-16(2)^2 + 64(2) + 60 = 124$
3	$-16(3)^2 + 64(3) + 60 = 108$
4	$-16(4)^2 + 64(4) + 60 = 60$

Plot the ordered pairs $(t, h(t))$ on the coordinate plane. Then connect the points.

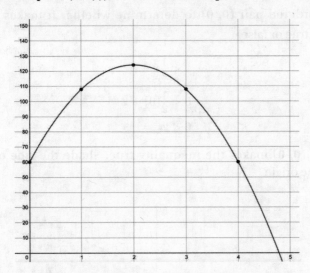

The average rate of change is calculated by dividing the change in height by the change in time. When Michael released the ball, the coordinates were $(0, 60)$. When the ball reached its maximum height, the coordinates were $(2, 124)$.

The average rate of change is calculated as follows:

$$\frac{124 - 60}{2 - 0} = \frac{64}{2} = 32$$

The average rate of change is 32 feet per second.

34. Rewrite both inequalities in slope-intercept form.

$$-x + 2y - 4 < 0$$
$$2y < x + 4$$
$$y < \frac{1}{2}x + 2$$

$$3x + 4y + 4 \geq 0$$
$$4y \geq -3x - 4$$
$$y \geq -\frac{3}{4}x - 1$$

Graph the line $y = \frac{1}{2}x + 2$ with a dotted line because the inequality is "less than" rather than "less than or equal to." Use the fact that the y-intercept is 2 and the slope is $\frac{1}{2}$ to help with the graphing.

Test the ordered pair (0, 0) to determine whether it makes the inequality $y < \frac{1}{2}x + 2$ true or false.

$$y \overset{?}{<} \frac{1}{2}x + 2$$
$$0 \overset{?}{<} \frac{1}{2}(0) + 2$$
$$0 < 2$$

The point (0, 0) makes the inequality true. Shade the side of the line that includes the origin.

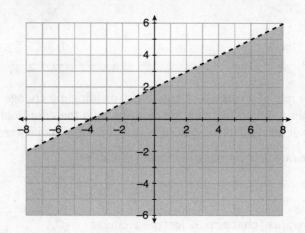

Graph the line $y = -\frac{3}{4}x - 1$ with a solid line because the inequality is "greater than or equal to" rather than "greater than." Use the fact that the y-intercept is -1 and the slope is $-\frac{3}{4}$ to help.

Test the ordered pair $(0, 0)$ to determine whether it makes the inequality $y \geq -\frac{3}{4}x - 1$ true or false.

$$y \overset{?}{\geq} -\frac{3}{4}x - 1$$

$$0 \overset{?}{\geq} -\frac{3}{4}(0) - 1$$

$$0 \geq -1$$

The point $(0, 0)$ makes the inequality true. Shade the side of the line that includes the origin.

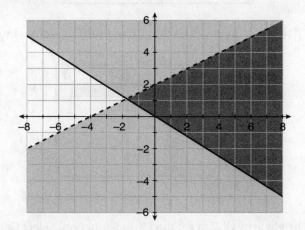

Notice that the ordered pair $(0, 0)$ is in the portion of the graph that is shaded in dark gray. It is in the overlap of the two inequalities. Therefore, $(0, 0)$ is a solution to this system. Stephen is correct.

35. A linear regression equation, or line of best fit, for a data set can be calculated with a graphing calculator.

For the TI-84:

To turn diagnostics on, press [2nd][0], scroll down to select DiagnosticOn, an press [Enter].

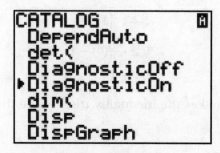

Press [STAT][1]. Enter the p-values into L1 and the $f(p)$-values into L2.

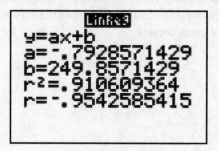

Press [STAT][right arrow][4] to get the LinReg function. Select Calculate.

For the TI-Nspire:

From the home screen, select the Add Lists & Spreadsheet icon. In the first row, label the A column "saleprice" and the B column "number." Enter the sale prices in cells A1 to A7. Enter the number of new homes available in cells B1 to B7.

	1.1	1.2	1.3	▶	*Unsaved ▼		
A	saleprice	B number	C		D		
◆							
1	160	126					
2	180	103					
3	200	82					
4	220	75					
5	240	82					
C1						◀ ▶	

Move the cursor to cell C1. Press [menu][4][1][3]. Set X List to "saleprice" and Y List to "number." Select OK.

	1.1	1.2	1.3	▶	*Unsaved ▼		
	B	number	C		D	E	
◆						=LinRegM	
1		126		Title		Linear Re..	
2		103		RegEqn		m*x+b	
3		82		m		-0.792857	
4		75		b		249.857	
5		82		r²		0.910609	
E1	="Linear Regression (mx+b)"					◀ ▶	

The line of best fit is $f(p) = -0.79p + 249.86$.

The correlation coefficient (r) is a measure of how close the points are to the lin of best fit. If the line of best fit has a positive slope, the correlation coefficier will be positive. If the line of best fit has a negative slope, the correlation coeff cient will be negative. The closer the correlation coefficient is to 1 (or to -1), th closer to the line of best fit the points will lie.

For this data set, the correlation coefficient is -0.95. Note that the calculatc computes both r and r^2. The question here asks for the value of r. Because th correlation coefficient is negative, the line of best fit must have a negative slope.

36. Let l be the length of the rectangle and w be the width of the rectangle, both in inches. According to the question, $l = \frac{1}{2}w + 6$. The area of a rectangle is given by the formula $A = lw$. If the area is 432 square inches:

$$432 = \left(\tfrac{1}{2}w + 6\right)(w)$$

Solve by writing the equation in standard form.

$$432 = \left(\tfrac{1}{2}w + 6\right)(w)$$

$$432 = \tfrac{1}{2}w^2 + 6w$$

$$0 = \tfrac{1}{2}w^2 + 6w - 432$$

There are many ways to solve this equation. Because some of the numbers ar large, it may be difficult to factor. The easiest way to proceed is to use the qua dratic formula.

$$w = \frac{-b \pm \sqrt{b^2 - 4ac}}{2a}$$

$$= \frac{-6 \pm \sqrt{(6)^2 - 4\left(\tfrac{1}{2}\right)(-432)}}{2\left(\tfrac{1}{2}\right)}$$

$$= -6 \pm \sqrt{(6)^2 - 4\left(\tfrac{1}{2}\right)(-432)}$$

$$= -6 \pm \sqrt{36 + 864}$$

$$= -6 \pm \sqrt{900}$$

$$= -6 \pm 30$$

This means that either $w = -36$ or $w = 24$. Since width cannot be negative, rejec the negative value. Therefore, the width is 24 inches.

The length is given by $l = \frac{1}{2}w + 6$.

$$l = \frac{1}{2}w + 6$$
$$= \frac{1}{2}(24) + 6$$
$$= 12 + 6$$
$$= 18$$

The length is 18 inches.

Part IV

37. The Brown family paid $170 for 3 children and 2 adults. That means that the sum of the children and adult tickets equals $170. An equation that represents this is $3x + 2y = 170$.

The Peckham family paid $360 for 4 children and 6 adults. An equation that represents this is $4x + 6y = 360$.

To graph the equations, convert to slope-intercept form.

$$3x + 2y = 170$$
$$2y = 170 - 3x$$
$$y = \frac{170 - 3x}{2}$$
$$y = -\frac{3}{2}x + 85$$

$$4x + 6y = 360$$
$$6y = 360 - 4x$$
$$y = \frac{360 - 4x}{6}$$
$$y = -\frac{2}{3}x + 60$$

Graph the lines on the grid.

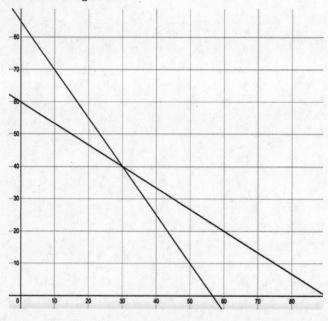

The graph shows that the point of intersection occurs at (30, 40).

The *x*-coordinate of the point of intersection represents the price of a child's ticket in dollars. Therefore, the price of each child's ticket was $30.

The *y*-coordinate of the point of intersection represents the price of an adult's ticket in dollars. Therefore, the price of each adult's ticket was $40.

Topic	Question Numbers	Number of Points	Your Points	Your Percentage
1. Polynomials	24, 28	$2 + 2 = 4$		
2. Properties of Algebra	5, 29, 32	$2 + 2 + 2 = 6$		
3. Functions	4, 13, 18	$2 + 2 + 2 = 6$		
4. Creating and Interpreting Equations	2, 14, 17, 21	$2 + 2 + 2 + 2 = 8$		
5. Inequalities	3, 26, 34	$2 + 2 + 4 = 8$		
6. Sequences and Series	8, 23	$2 + 2 = 4$		
7. Systems of Equations	20, 37	$2 + 6 = 8$		
8. Quadratic Equations and Factoring	6, 9, 12, 15, 16, 31, 36	$2 + 2 + 2 + 2 + 2 + 2 + 4 = 16$		
9. Regression	35	4		
10. Exponential Equations	1, 19	$2 + 2 = 4$		
11. Graphing	7, 11, 25, 33	$2 + 2 + 2 + 4 = 10$		
12. Statistics	10, 22	$2 + 2 = 4$		
13. Number Properties	30	2		
14. Unit Conversions	27	2		

How to Convert Your Raw Score to Your Algebra I Regents Exam Score

The accompanying conversion chart must be used to determine your final score on the January 2020 Regents Exam in Algebra I. To find your final exam score, locate in the column labeled "Raw Score" the total number of points you scored out of a possible 86 points. Since partial credit is allowed in Parts II, III, and IV of the test, you may need to approximate the credit you would receive for a solution that is not completely correct. Then locate in the adjacent column to the right the scale score that corresponds to your raw score. The scale score is your final Algebra I Regents Exam score.

Regents Exam in Algebra I—January 2020
Chart for Converting Total Test Raw Scores
to Final Exam Scores (Scale Scores)

Raw Score	Scale Score	Performance Level	Raw Score	Scale Score	Performance Level	Raw Score	Scale Score	Performance Level
86	100	5	57	81	4	28	66	3
85	99	5	56	81	4	27	65	3
84	97	5	55	81	4	26	64	2
83	96	5	54	80	4	25	63	2
82	95	5	53	80	4	24	61	2
81	94	5	52	80	4	23	60	2
80	93	5	51	80	4	22	59	2
79	92	5	50	79	3	21	57	2
78	91	5	49	79	3	20	56	2
77	90	5	48	79	3	19	55	2
76	89	5	47	78	3	18	52	1
75	89	5	46	78	3	17	50	1
74	88	5	45	78	3	16	48	1
73	87	5	44	77	3	15	46	1
72	87	5	43	77	3	14	44	1
71	86	5	42	76	3	13	42	1
70	86	5	41	76	3	12	40	1
69	86	5	40	75	3	11	37	1
68	85	5	39	75	3	10	35	1
67	84	4	38	74	3	9	32	1
66	84	4	37	74	3	8	29	1
65	84	4	36	73	3	7	26	1
64	83	4	35	72	3	6	23	1
63	83	4	34	72	3	5	20	1
62	83	4	33	71	3	4	16	1
61	82	4	32	70	3	3	12	1
60	82	4	31	69	3	2	9	1
59	82	4	30	68	3	1	4	1
58	81	4	29	67	3	0	0	1

Sample 2024 Exam
Algebra I

High School Math Reference Sheet

Conversions

1 inch = 2.54 centimeters

1 meter = 39.37 inches

1 mile = 5280 feet

1 mile = 1760 yards

1 mile = 1.609 kilometers

1 kilometer = 0.62 mile

1 pound = 16 ounces

1 pound = 0.454 kilogram

1 kilogram = 2.2 pounds

1 ton = 2000 pounds

1 cup = 8 fluid ounces

1 pint = 2 cups

1 quart = 2 pints

1 gallon = 4 quarts

1 gallon = 3.785 liters

1 liter = 0.264 gallon

1 liter = 1000 cubic centimeters

Formulas

Triangle	$A = \frac{1}{2}bh$
Parallelogram	$A = bh$
Circle	$A = \pi r^2$
Circle	$C = \pi d \text{ or } C = 2\pi r$
General Prisms	$V = Bh$
Cylinder	$V = \pi r^2 h$
Sphere	$V = \frac{4}{3}\pi r^3$

Formulas (continued)

Cone	$V = \frac{1}{3}\pi r^2 h$
Pyramid	$V = \frac{1}{3}Bh$
Pythagorean Theorem	$a^2 + b^2 = c^2$
Quadratic Formula	$x = \dfrac{-b \pm \sqrt{b^2 - 4ac}}{2a}$
Arithmetic Sequence	$a_n = a_1 + (n-1)d$
Geometric Sequence	$a_n = a_1 r^{n-1}$
Geometric Series	$S_n = \dfrac{a_1 - a_1 r^n}{1 - r}$ where $r \neq 1$
Radians	$1 \text{ radian} = \frac{180}{\pi} \text{ degrees}$
Degrees	$1 \text{ degree} = \frac{\pi}{180} \text{ radians}$
Exponential Growth/Decay	$A = A_0 e^{k(t - t_0)} + B_0$

PART I

Answer all 24 questions in this part. Each correct answer will receive 2 credits. No partial credit will be allowed. For each statement or question, write in the space provided the numeral preceding the word or expression that best completes the statement or answers the question.   [48 credits]

1. Below is the graph of the solution set of the equation $3x + 2y = 36$.
   Which point is *not* on this line?

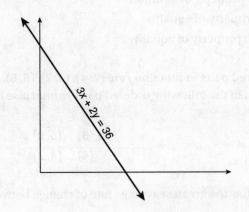

$$\frac{36 - 3x}{2}$$

    (1) $(2, 15)$ (3) $(9, 6)$

    (2) $(4, 12)$ (4) $(10, 3)$ 1 _____

2. $f(x) = x^2$ and $g(x) = 6x - 9$. For what value of x does $f(x) = g(x)$?

    (1) -3 (3) -9

    (2) 3 (4) 9 2 _____

3. Which expression is equal to the 99th term of the sequence
   3, 10, 17, 24, ...?

    (1) $3 + 7 \times 98$ (3) $3 + 7 \times 100$

    (2) $3 + 7 \times 99$ (4) $7 + 3 \times 98$ 3 _____

4. To find the solution to the equation $3x + 5 = 14$, a student does the following first step:

$$3x + 5 = 14$$
$$-5 = -5$$
$$3x = 9$$

Which property of algebra can be used to justify this first step?

(1) distributive property

(2) associative property of addition

(3) addition property of equality

(4) subtraction property of equality

4 _____

5. Four of the ordered pairs in function f are $(2, 4)$, $(3, 7)$, $(5, 6)$, and $(9, 1)$. Which of the following ordered pairs *cannot* also be in function f?

(1) $(7, 6)$ (3) $(5, 8)$

(2) $(4, 8)$ (4) $(4, 2)$

5 _____

6. Which function has the greatest average rate of change between $x = 1$ and $x = 3$?

(1) $f(x) = 2^x$ (3) $f(x) = 2x$

(2) $f(x) = x^2$ (4) $f(x) = |3x| + 1$

6 _____

7. If $A = \sqrt{2}$ and $B = \dfrac{2}{\sqrt{2}}$, which of the following is a rational number?

(1) A (3) $A + B$

(2) B (4) $A \times B$

7 _____

8. Which data set has the smallest sample standard deviation?

Data set A: 5, 6, 8, 8, 9, 12

Data set B: 4, 5, 8, 9, 11, 14

Data set C: 7, 8, 8, 8, 8, 9

(1) Data set A

(2) Data set B

(3) Data set C

(4) They all have the same sample standard deviation.

8 _____

9. You are interested in calculating how many gallons of gasoline you would need to purchase to fill up the gas tank of your car. The appropriate domain for the number of gallons of gasoline needed would consist of

(1) rational numbers

(2) integers

(3) positive rational numbers

(4) positive integers   9 _____

10. Which of the following is *not* equivalent to $4 \cdot 4^{2x}$?

(1) 16^{2x} (3) 2^{4x+2}

(2) 4^{2x+1} (4) $4 \cdot 16^x$ 10 _____

11. Twenty adults take a math test. A scatterplot of the data is created. The *x*-axis displays the day of the month the test takers were born on, and the *y*-axis displays the score they achieved on the math test. Below is that scatterplot based on the data collected.

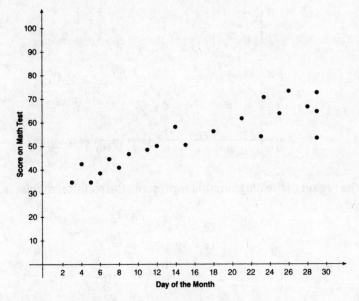

Which statement about these data is most accurate?

(1) There is a causation in the data but not a correlation.

(2) There is a correlation in the data but not a causation.

(3) There is both a causation and a correlation in the data.

(4) There is neither a causation nor a correlation in the data.   11 _____

12. What are the zeros of the function $f(x) = x^2 - 5x$?

    (1) $\{5\}$ x -5   (3) $\{-5\}$

    (2) $\{0, 5\}$ (4) $\{0, -5\}$ 12 _____

13. The value of a rare baseball card over time, in years, can be calculated with the formula $V = 5 \cdot 1.02^t$. What does the number 1.02 represent in this formula?   13 __2__

    (1)   the percent increase in 1 year

    (2)   1 more than the percent increase in 1 year

    (3)   1 less than the percent increase in 1 year

    (4)   the starting value of the card

14. The dashed lines in the following diagram represent $4x + 3y = 24$ and $3x + 5y = 30$.

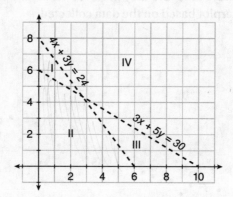

Select the region of the diagram that represents the solution to this inequality.

$$4x + 3y < 24$$
$$3x + 5y > 30$$

    (1)   I   (3)   III

    (2)   II   (4)   IV   14 _____

15. This is the graph of the function $f(x) = x^2$.

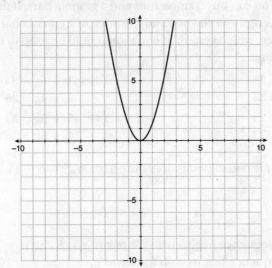

Which of these functions has a minimum at the point $(5, -2)$?

(1) $g(x) = f(x+5) - 2$

(2) $g(x) = f(x-5) - 2$ 2

(3) $g(x) = f(x-5) + 2$

(4) $g(x) = f(x+5) + 2$ 15_____

16. A social media video had 100 views on the first day it was posted and 300 views by the next day. If the growth rate is exponential, how many views will the video have on the fourth day?

(1)   500   (3)   900

(2)   700   (4)   2700   16_____

17. Fully simplify the expression $(x-3)^2 - (x+3)(x-3)$.

$x^2 - 6x + 9 - x^2 - 3x - 9 + 3x$

$(x-3)(x-3)$

(1) $-6x$ $-6x + 9 - 3x - 9 + 3x$ (3) 0 $x^2 - 3x - 3x + 9$

(2) $-6x + 18$ (4) $-6x - 18$ 17_____

18. The graph of which of these functions has vertical symmetry?

(1) $f(x) = x$ (3) $f(x) = -x$ 2

(2) $f(x) = 2^x$ (4) $f(x) = x^2$ 18_____

19. At a health food store, for \$25, you can buy 3 smoothies and 5 granola bars. For \$33, you can buy 5 smoothies and 4 granola bars. If the price of 1 smoothie is represented by s and the price of 1 granola bar is represented by g, which system of equations could be used to find the values of s and g?

(1) $3s + 5g = 25$
     $5s + 4g = 33$

(3) $3s + 5g = 33$
     $5s + 4g = 25$

(2) $5s + 3g = 25$
     $4s + 5g = 33$

(4) $5s + 3g = 33$
     $4s + 5g = 25$

19 ___2___

20. How many unique real solutions does the equation $x^2 + 6x + 15 = 0$ have?

(1) 1

(2) 2

(3) 3

(4) 0

20 ___2___

21. An arithmetic sequence begins with the terms $a_1 = 5$ and $a_2 = 10$. A geometric sequence begins with the terms $g_1 = 5$ and $g_2 = 10$. What is the value of $g_5 - a_5$?

(1) 55

(2) 0

(3) −55

(4) 20

21 ___2___

22. Function $f(x)$ is represented by the following chart, and function $g(x)$ is represented by the graph. What is the value of $f(3) + g(2)$?

x	f(x)
1	8
2	4
3	7
4	6
5	3

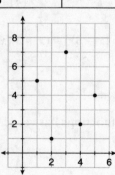

(1) 11

(2) 9

(3) 5

(4) 8

22 ___γ___

23. Eleven data points are plotted from the chart below to form a scatterplot.

x	y
1	1
1	2
2	1
2	2
3	2
4	2
4	3
5	2
5	3
6	3
6	4

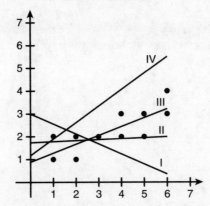

For this scatterplot, which line has the greatest r^2-value?

(1)   I   (3)   III

(2)   II   (4)   IV   23 _____

24. The price you pay for an app, including n in-app purchases, can be calculated with the formula $P = 3.99 + 0.99n$. What does the number 0.99 represent?

(1)   the total price of the app with the in-app purchases

(2)   the price of the app without any in-app purchases

(3)   the price of one in-app purchase

(4)   the difference between the price of the app and the price of one in-app purchase   24 _____

PART II

Answer all 8 questions in this part. Each correct answer will receive 2 credits. Clearly indicate the necessary steps, including appropriate formula substitutions, diagrams, graphs, charts, etc. For all questions in this part, a correct numerical answer with no work shown will receive only 1 credit.   [16 credits]

25. Rewrite the polynomial $5x + 3x^2 + 1$ in standard form.

26. For a birthday party, a family has $100 to spend. If there are c children and a adults and if it costs $8 to feed an adult and $5 to feed a child, what inequality represents the possible values of c and a?

27. Write $\frac{5}{\sqrt{3}}$ as an expression with a rational denominator.

28. Solve for x:

$$\frac{2}{3}x + 10 = 6$$

29. What are all the roots of the equation $(x^2 - 9)(x - 2)(x + 5) = 0$?

30. The equation that relates the amount of money (A) you would have in the bank after 2 years, if you get an interest rate of r and initially put P into the bank, is $A = P(1 + r)^2$. Solve this equation for r in terms of A and P.

31. Three hundred students are asked the following two questions on a survey: (1) Do you take art or music class? and (2) Do you prefer science or social studies? Some of the results are displayed in the two-way frequency table below.

	Art	Music	Total
Science	50		
Social Studies		75	
Total	135	165	300

What is the probability that a student prefers social studies? Round your answer to the *nearest tenth of a percent*.

What is the probability that a student prefers social studies if it is known that the student takes art class? Round your answer to the *nearest tenth of a percent*.

32. Below is a graph of the following system of equations:

$$-2x + 3y = 12$$
$$4x + 5y = 42$$

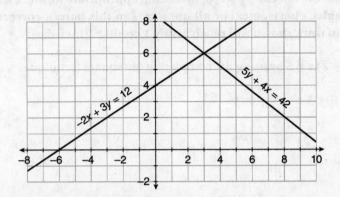

Use the graph or any algebraic method to find the solution to this system of equations. Explain how you arrived at your answer.

PART III

Answer all 4 questions in this part. Each correct answer will receive 4 credits. Clearly indicate the necessary steps, including appropriate formula substitutions, diagrams, graphs, charts, etc. For all questions in this part, a correct numerical answer with no work shown will receive only 1 credit. [16 credits]

33. Solve $x^2 - 24x = 5$ by using the completing the square technique.

34. For what values of x does $(x + 5)^2 + 4(x + 5) - 12 = 0$? Show all your work, including an algebraic solution.

35. The ages of the eleven people sitting in the front row of a concert were 12, 35, 13, 9, 41, 14, 39, 11, 41, 12, 65. Find the median, first quartile, and third quartile for this data set, and use those values to create a box plot of the data.

36. Sketch the graphs of $f(x) = 2 \cdot 3^x$ and $g(x) = 10$ on the interval $-4 \leq x \leq 2$.

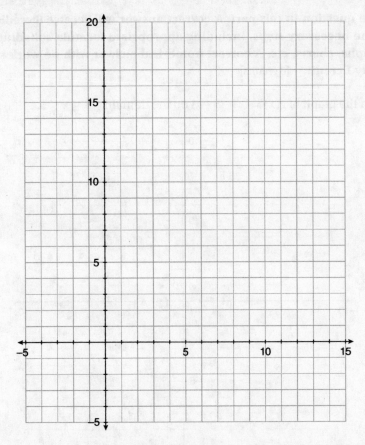

Using this graph, determine the two integers between which $2 \cdot 3^x = 10$.

PART IV

Answer the question in this part. A correct answer will receive 6 credits. Clearly indicate the necessary steps, including appropriate formula substitutions, diagrams, graphs, charts, etc. A correct numerical answer with no work shown will receive only 1 credit.   [6 credits]

37. Sketch the graph of $g(x) = |x + 2| - 3$ on the domain $-3 \le x \le 5$.

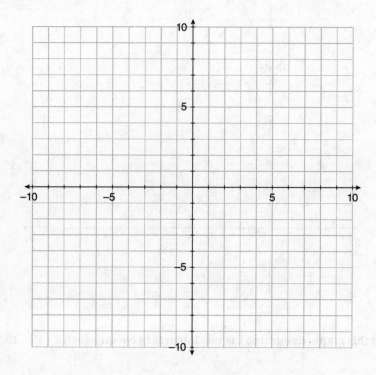

Answers Sample 2024 Exam
Algebra I

Answer Key

PART I

1. (3)	5. (3)	9. (3)	13. (2)	17. (2)	21. (1)
2. (2)	6. (2)	10. (1)	14. (1)	18. (4)	22. (4)
3. (1)	7. (4)	11. (2)	15. (2)	19. (1)	23. (3)
4. (4)	8. (3)	12. (2)	16. (4)	20. (4)	24. (3)

PART II

25. $3x^2 + 5x + 1$

26. $5c + 8a \le 100$ or $8a + 5c \le 100$

27. $\dfrac{5\sqrt{3}}{3}$

28. $x = -6$

29. $x = 3, x = -3, x = 2$, and $x = -5$

30. $\sqrt{\dfrac{A}{P}} - 1 = r$

31. 53.3%; 63.0%

32. $x = 3, y = 6$, or $(3, 6)$

PART III

33. $12 \pm \sqrt{149}$

34. $\{-3, -11\}$

35. Median = 14; first quartile = 12; third quartile = 41;

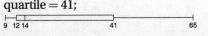

36.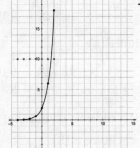
$x = 1$ and $x = 2$

PART IV

37.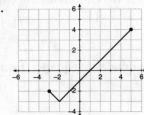

In **Parts II–IV**, you are required to show how you arrived at your answers. For sample methods of solutions, see the *Answer Explanations* section.

Answer Explanations

Part I

1. Since there are no gridlines on this graph, the best way to tell if a point is on the line is to plug the x-coordinate for x and the y-coordinate for y and then see which of these ordered pairs does not lead to a true expression.

    Testing choice (1):

    $$3x + 2y = 3(2) + 2(15) = 6 + 30 = 36$$
    $$36 = 36$$

    Testing choice (2):

    $$3x + 2y = 3(4) + 2(12) = 12 + 24 = 36$$
    $$36 = 36$$

    Testing choice (3):

    $$3x + 2y = 3(9) + 2(6) = 27 + 12 = 39$$

    Testing choice (4):

    $$3x + 2y = 3(10) + 2(3) = 30 + 6 = 36$$
    $$39 \neq 36$$

    Choice (3) is the only choice that does not equal 36.

    The correct choice is **(3)**.

2. To find the x-value for which the functions are equal, create an equation where the function definitions are set equal to one another.

    $$f(x) = g(x)$$
    $$x^2 = 6x - 9$$

    Solve this quadratic equation by moving all the terms to the left-hand side of the equal sign.

    $$x^2 - 6x + 9 = 0$$

There are several ways of solving for x: factoring, completing the square, the quadratic formula, or graphing. The fastest way is to use factoring.

$$x^2 - 6x + 9 = 0$$
$$(x - 3)(x - 3) = 0$$
$$x - 3 = 0$$
$$x = 3$$

The correct choice is **(2)**.

3. The sequence 3, 10, 17, 24, . . . is an arithmetic sequence because each term is equal to the previous term plus or minus the same number each time. In this sequence, each term is equal to the previous term plus 7. To find the 99th term, use the arithmetic sequence formula from the reference sheet:

$$a_n = a_1 + (n - 1)d$$

where a_1 is the first term of the sequence, d is the number you have to add to a term in order to get to the next term, and n is the position in the sequence of the term you are trying to calculate.

$$a_{99} = 3 + (99 - 1)7 = 3 + 98 \times 7$$

Since the answer choices do not require you to calculate the actual value of the term, see if any of the answer choices are equivalent to this expression.

The correct choice is **(1)**.

4. When you perform the same operation on both sides of an equation, the solution of the equation does not change. In this step, the student eliminated the 5 from the left side of the equation by subtracting 5 from both sides. When you subtract the same number from both sides of an equation, you are using the subtraction property of equality.

Choice (1) is not correct. The distributive property is when you multiply a value by an expression with multiple terms in parentheses. For example, $5(2x + 3) = 10x + 15$ is justified by the distributive property.

Choice (2) is not correct. The associative property of addition is when you group a sum or product in a different way. For example, $5x + (2x + 3) = (5x + 2x) + 3$ is justified by the associative property of addition.

Choice (3) is not correct. The addition property of equality is used when you add the same value to both sides of an equation. For example, if the equation was $3x - 5 = 14$ and the student eliminated the -5 by adding 5 to both sides of the equation, this would be justified by the addition property of equality.

The correct choice is **(4)**.

5. In a function, all the x-coordinates of the ordered pairs must be different.

Testing choice (1): $(7, 6)$ can be in the function since there is no other point with an x-coordinate of 7. It is not a problem that there is another point, $(5, 6)$, with a y-coordinate of 6.

Testing choice (2): $(4, 8)$ can be in the function since there is no other point with an x-coordinate of 4.

Testing choice (3): $(5, 8)$ *cannot* be in the function since there is already a point, $(5, 6)$, with an x-coordinate of 5.

Testing choice (4): $(4, 2)$ can be in the function since there is no other point with an x-coordinate of 4.

The correct choice is **(3)**.

6. The average rate of change of a function from $x = a$ to $x = b$ can be calculated with this formula:

$$\text{Average rate of change} = \frac{f(b) - f(a)}{b - a}$$

Testing choice (1):

$$\frac{2^3 - 2^1}{3 - 1} = \frac{8 - 2}{3 - 1}$$
$$= \frac{6}{2}$$
$$= 3$$

Testing choice (2):

$$\frac{3^2 - 1^2}{3 - 1} = \frac{9 - 1}{3 - 1}$$
$$= \frac{8}{2}$$
$$= 4$$

Testing choice (3):

$$\frac{2(3) - 2(1)}{3 - 1} = \frac{6 - 2}{3 - 1}$$
$$= \frac{4}{2}$$
$$= 2$$

Testing choice (4):

$$\frac{(|3 \cdot 3| + 1) - (|3 \cdot 1| + 1)}{3 - 1} = \frac{10 - 4}{3 - 1}$$
$$= \frac{6}{2}$$
$$= 3$$

The correct choice is **(2)**.

7. A rational number is one that can be expressed as a fraction that has an integer in both the numerator and the denominator. $\frac{3}{4}$ and $\frac{1}{7}$ are two examples of rational numbers. When a rational number is converted into a decimal, it will either terminate, like $\frac{3}{4} = 0.75$, or the decimal will have a repeating pattern, like $\frac{1}{7} = 0.142857142857\ldots$. An irrational number will neither repeat nor terminate.

Testing choice (1): If you enter $\sqrt{2}$ into a calculator, you will get $1.41421356237\ldots$. So $\sqrt{2}$ is an irrational number as is the square root of any number that is not a perfect square.

Testing choice (2): If you enter $\frac{2}{\sqrt{2}}$ into a calculator, you will also get $1.41421356237\ldots$. Generally, if you divide a rational number by an irrational number, the quotient will be an irrational number.

Testing choice (3): Although it is possible to add two irrational numbers and get a rational sum, in this case if you add the decimals from choice (1) and choice (2), you will get $2.82842712474\ldots$, which is irrational. If you notice that the numbers from choice (1) and choice (2) are equivalent, this sum can be thought of as $\sqrt{2} + \sqrt{2} = 2\sqrt{2}$, and generally a rational number multiplied by an irrational number will be irrational.

Testing choice (4): Usually, when you multiply two irrational numbers, you will get an irrational answer. However, it is possible for the product to be rational. In this example, if you multiply the values for A and B on a calculator, you will get 2, which is rational. You could also see that this will happen even without a calculator since the $\sqrt{2}$ in the numerator of the first expression will cancel out the $\sqrt{2}$ in the denominator of the second expression.

The correct choice is **(4)**.

8. The sample standard deviation is a measure of how close the numbers in a sample data set are to each other. The closer the sample standard deviation is to zero, the closer together the numbers are. The sample standard deviation can be calculated on a graphing calculator using the following steps, which outline how to calculate the sample standard deviation for data set A.

On the TI-84:

Press [STAT] [1] to enter the numbers into L1.

L1(7)=

Press [STAT], go to CALC menu, and press [1] for 1-Variable Statistics.

```
EDIT CALC TESTS
1 1-Var Stats
2:2-Var Stats
3:Med-Med
4:LinReg(ax+b)
5:QuadReg
6:CubicReg
7↓QuartReg
```

Set List to L1 and leave FreqList blank. Then highlight Calculate and press [ENTER].

```
1-Var Stats
List:L1
FreqList:
Calculate
```

```
1-Var Stats
x̄=8
Σx=48
Σx²=414
Sx=2.449489743
σx=2.236067977
↓n=6
```

The sample standard deviation is denoted by Sx, which is approximately 2.4 for data set A. Underneath that value is $\sigma_x = 2.2$, which is the population standard deviation.

On the TI-Nspire:

From the home screen, select the Add Lists & Spreadsheet icon.

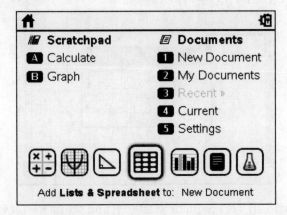

Name column A "x," and fill in cells A1 through A6 with the 6 numbers from data set A.

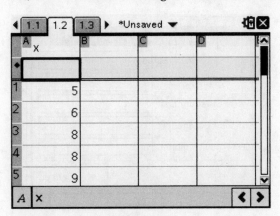

Press [menu], [4], [1], and [1] for One-Variable Statistics.

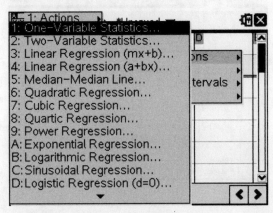

Select the [OK] button.

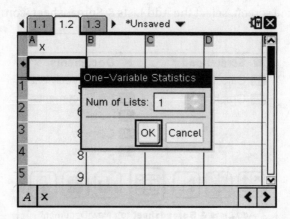

Set the X1 List to a[], set Frequency List to 1, and set 1st Result Column to b[].
Then select [OK].

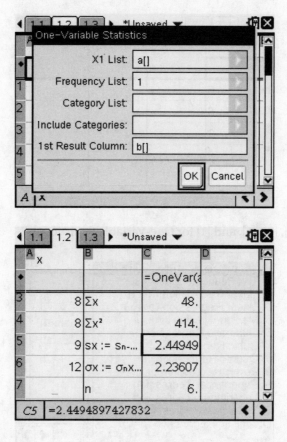

The sample standard deviation is denoted by *Sx*, which is approximately 2.4 for data set A.

When you calculate the sample standard deviation for the other two data sets (using the same steps outlined for data set A), you discover that the sample standard deviation is 3.7 for data set B and 0.6 for data set C. Data set C has the smallest sample standard deviation.

This question can also be solved without using a graphing calculator. Knowing that a data set with very similar numbers has a low standard deviation and seeing how only one of these data sets has numbers that are very close together (or even the same number repeated), you can infer that data set C has the lowest sample standard deviation. Still, it's worthwhile to check with the calculator if there is no data set that has numbers so close together.

The correct choice is **(3)**.

9. Since you cannot buy a negative amount of gasoline, choices (1) and (2) can be eliminated because they include negative numbers. Gasoline does not come in only 1-gallon units. You can buy fractions of a gallon, like $5\frac{1}{2}$ gallons, which is a rational number but not an integer. Therefore, the most appropriate domain would consist of positive rational numbers.

The correct choice is **(3)**.

10. One way to simplify this expression is to write it as $4^1 \cdot 4^{2x}$ and use the property of exponents that $a^x \cdot a^y = a^{x+y}$.

$$4^1 \cdot 4^{2x} = 4^{1+2x} = 4^{2x+1}$$

This is choice (2), so choice (2) is an equivalent expression.

Another way to simplify this expression is to use the property of exponents that $a^{xy} = (a^x)^y$.

$$4 \cdot 4^{2x} = 4 \cdot (4^2)^x = 4 \cdot 16^x$$

This is choice (4), so choice (4) is an equivalent expression.

A third way to simplify this expression is to use the fact that $4 = 2^2$ and rewrite the original expression as follows.

$$2^2 \cdot (2^2)^{2x} = 2^2 \cdot 2^{4x} = 2^{2+4x} = 2^{4x+2}$$

This is choice (3), so choice (3) is an equivalent expression.

Therefore, the only one that is *not* equivalent is choice (1).

Another way to determine which expression is not equivalent is to plug a number, like $x = 3$, into the original expression and into each of the choices to see which choice has a different value.

When $x = 3$, the original expression becomes $4 \cdot 4^{2 \cdot 3} = 4 \cdot 4^6 = 4 \cdot 4096 = 16{,}384$. When you substitute the same value, $x = 3$, into choice (1), it becomes $16^{2 \cdot 3} = 16^6 = 16{,}777{,}216$.

The correct choice is **(1)**.

11. Because the points in the plot do resemble a line, there is a correlation in the data. However, having a correlation does not mean that one of the things being measured has in any way caused or affected the results of the other thing being measured. It is very unlikely that being born later in a month would cause someone to score higher on a math test. Therefore, even though there is a correlation, this does not seem to be a causal relationship.

In general, when there is a causation, there will also be a correlation. However, when there is a correlation, there does not have to be causation.

The correct choice is **(2)**.

12. A zero of a function is an x-value that makes $f(x) = 0$.

$$f(x) = x^2 - 5x = 0$$
$$x^2 - 5x = 0$$

Factor out the common factor of x.

$$x(x - 5) = 0$$

This is true if $x = 0$ or if $x - 5 = 0$, so the two zeros are $x = 0$ and $x = 5$.

Since this question is multiple-choice, you could also test the numbers in the answer choices to see which make $f(x) = 0$.

Testing 5:

$$f(5) = 5^2 - 5(5) = 25 - 25 = 0$$

Testing 0:

$$f(0) = 0^2 - 5(0) = 0 - 0 = 0$$

Testing -5:

$$f(-5) = (-5)^2 - 5(-5) = 25 + 25 = 50$$

Of the numbers in the choices, the only ones that work are $x = 5$ and $x = 0$.

The correct choice is **(2)**.

13. The formula for annual exponential growth is $V = P(1 + r)^t$, where P is the starting value, t is the time in years, and r is the percent increase. For this situation, the formula can be written as $V = P(1 + 0.02)^t$, so the percent increase is 0.02 or 2%. The 1.02 therefore represents $1 + 0.02$, which is 1 more than the percent increase in 1 year.

The correct choice is **(2)**.

14. One way to answer this question is to test a point in each region and see if it makes both inequalities true.

Testing choice (1): $(1, 6)$ is in region I. Substituting $x = 1$ and $y = 6$ into the first inequality becomes:

$$4x + 3y < 24$$
$$4(1) + 3(6) \overset{?}{<} 24$$
$$4 + 18 \overset{?}{<} 24$$
$$22 \overset{\checkmark}{<} 24$$

This is a true statement. In order for this to be the correct region, the other inequality must hold true for the same point. Substituting $x = 1$ and $y = 6$ into the second inequality becomes:

$$3x + 5y > 30$$
$$3(1) + 5(6) \overset{?}{>} 30$$
$$3 + 30 \overset{?}{>} 30$$
$$33 \overset{\checkmark}{>} 30$$

Since this is also a true statement, the answer must be region I.

Another way to solve this problem is to shade the proper side of each line and see where the double-shaded region is. When you test if $(0, 0)$ is in the first inequality, it becomes:

$$4x + 3y < 24$$
$$4(0) + 3(0) \overset{?}{<} 24$$
$$0 + 0 \overset{?}{<} 24$$
$$0 \overset{\checkmark}{<} 24$$

This is a true statement, so shade the side of the steeper line that contains (0, 0).

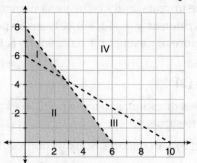

For the other inequality, when you test (0, 0), it becomes:

$$3x + 5y > 30$$
$$3(0) + 5(0) \overset{?}{>} 30$$
$$0 + 0 \overset{?}{>} 30$$
$$0 \not> 30$$

Since this is not a true statement, shade the side of the less steep line that does *not* contain (0, 0).

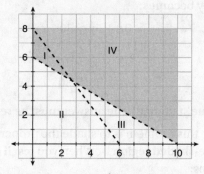

The region that was shaded in both diagrams is region I.

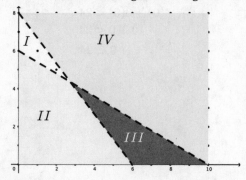

The correct choice is **(1)**.

15. The original function, $f(x)$, has a minimum at the point $(0, 0)$. For the new function to have a minimum at the point $(5, -2)$, the graph of the original function has to be translated 5 units to the right and 2 units down.

The graph of a function of the form $f(x - h) + k$ is like the graph of $f(x)$ but shifted h units to the right and k units up. In this case, $h = 5$ and, because the graph is shifting down instead of up, $k = -2$. So, the solution is $g(x) = f(x - 5) - 2$.

This question can also be answered by graphing the four answer choices and seeing which of the new graphs has a minimum point at $(5, -2)$.

Testing choice (1):

$$g(x) = f(x + 5) - 2 = (x + 5)^2 - 2$$

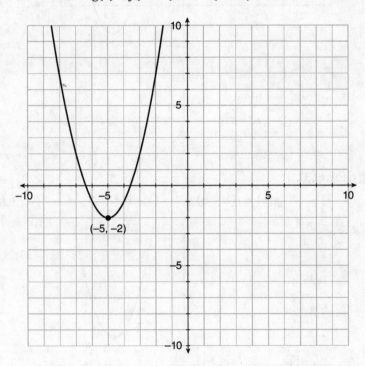

$(-5, -2)$

Testing choice (2):

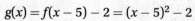

$$g(x) = f(x - 5) - 2 = (x - 5)^2 - 2$$

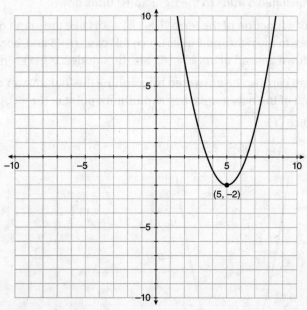

Testing choice (3):

$$g(x) = f(x - 5) + 2 = (x - 5)^2 + 2$$

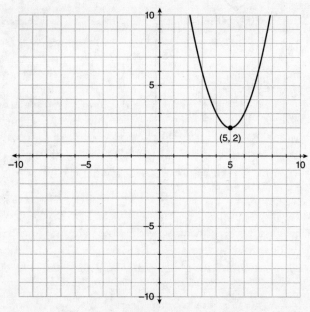

Testing choice (4):

$$g(x) = f(x+5) + 2 = (x+5)^2 + 2$$

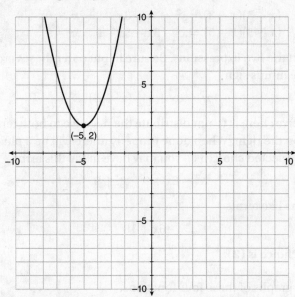

(−5, 2)

The correct choice is **(2)**.

16. A geometric sequence is one in which each term is equal to the previous term multiplied by the same number each time. From the reference sheet, you'll find the formula for the nth term of a geometric sequence:

$$a_n = a_1 r^{n-1}$$

where r is what you multiply each term by to get the next term and a_1 is the first term of the sequence.

To get the fourth term of this sequence, use $a_1 = 100$, $r = 3$, and $n = 4$:

$$\begin{aligned}
a_4 &= 100 \cdot 3^{4-1} \\
&= 100 \cdot 3^3 \\
&= 100 \cdot 27 \\
&= 2700
\end{aligned}$$

Since the value of n is so small, this question can also be answered without using the formula as long as you know that for a geometric sequence, each term is obtained from the previous term by multiplying by the same number. Since the first two terms of the sequence are 100 and 300 and since $300 = 3 \cdot 100$, the third term will be $3 \cdot 300 = 900$ and the fourth term will be $3 \cdot 900 = 2700$.

The correct choice is **(4)**.

17. Rewrite the expression as $(x - 3)(x - 3) - (x + 3)(x - 3)$.

Then simplify by multiplying the binomials.

$$(x - 3)(x - 3) = x^2 - 3x - 3x + 9 = x^2 - 6x + 9$$
$$(x + 3)(x - 3) = x^2 - 3x + 3x - 9 = x^2 - 9$$

Now subtract the two expressions. Be sure to put parentheses around the expression after the minus sign.

$$x^2 - 6x + 9 - (x^2 - 9)$$

A common error here is to not distribute the negative to both terms in the expression on the right. When done properly, there will be a plus sign before the last 9. If you selected choice (1), you may have made this error.

$$x^2 - 6x + 9 - x^2 + 9$$

Now combine like terms.

$$-6x + 18$$

The correct choice is **(2)**.

18. Here are the graphs of these four functions.

Choice (1):

$$f(x) = x$$

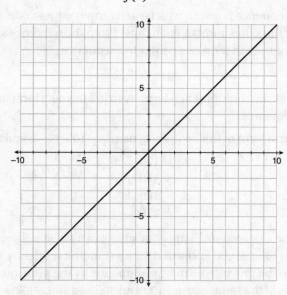

This graph has point symmetry and also has symmetry about the lines $y = x$ and $y = -x$. However, it does not have vertical symmetry.

Choice (2):

$$f(x) = 2^x$$

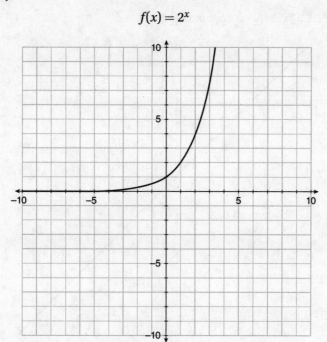

This graph does not have any symmetry.

Choice (3):

$$f(x) = -x$$

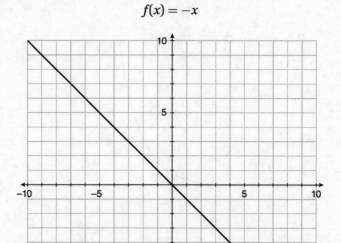

This graph has point symmetry and also has symmetry about the lines $y = x$ and $y = -x$. However, it does not have vertical symmetry.

Choice (4):

$$f(x) = x^2$$

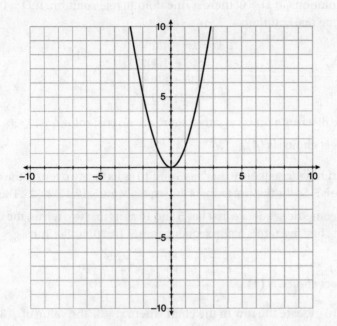

This graph has vertical symmetry. The vertical line of symmetry is the y-axis.

The correct choice is **(4)**.

19. If 1 smoothie costs s dollars, then 3 smoothies cost $3 \times s = 3s$.

    If 1 granola bar costs g dollars, then 5 granola bars cost $5 \times g = 5g$.

    If 3 smoothies and 5 granola bars cost $25, the equation is $3s + 5g = 25$.

    Similarly, if 5 smoothies and 4 granola bars cost $33, the equation is $5s + 4g = 33$.

    The correct choice is **(1)**.

20. To find the number of unique real solutions of a quadratic equation of the form $ax^2 + bx + c = 0$, first calculate the discriminant, $D = b^2 - 4ac$. If $D < 0$, there are no real solutions. If $D = 0$, there is one unique real solution. If $D > 0$, there are two unique real solutions.

$$\begin{aligned} D &= b^2 - 4ac \\ &= 6^2 - 4(1)(15) \\ &= 36 - 60 \\ &= -24 < 0 \end{aligned}$$

Since the discriminant is negative, there are no real solutions.

The correct choice is **(4)**.

21. For the arithmetic sequence that has 5 and 10 as its first two terms, the common difference is 5. So the first five terms of the sequence are 5, 10, 15, 20, 25. Thus, $a_5 = 25$.

For the geometric sequence that has 5 and 10 as its first two terms, the common ratio is 2. So the first five terms of the sequence are 5, 10, 20, 40, 80. Thus, $g_5 = 80$.

$$g_5 - a_5 = 80 - 25 = 55$$

The correct choice is **(1)**.

22. To find $f(3)$, locate the row in the chart where $x = 3$. The value of $f(3)$ is in the second column of that row, so $f(3) = 7$.

To find $g(2)$, locate the point on the graph that has an x-coordinate of 2. The value of $g(2)$ is the y-coordinate of that point, so $g(2) = 1$.

$$f(3) + g(2) = 7 + 1 = 8$$

The correct choice is **(4)**.

23. When a line comes very close to the points of a scatterplot, it is known as a good fit. There is one line that is considered the line of best fit. When one line comes closer to the points of a scatterplot than another line, it will have a higher r-value and a higher r^2-value. r is the correlation coefficient, and it can be as low as -1 or as high as $+1$. r^2 will always be positive and is a number between 0 and 1. The closer the line is to the points on the scatterplot, the closer r^2 is to 1.

For this question, it is not necessary to find the line of best fit since there is only one line, line III, that is a much better fit than the other three lines. Line III would have the highest r^2-value, followed by line IV, then line II, and finally line I.

By using a graphing calculator, you can see what the line of best fit is for a data set. This line would have a higher r^2-value than any other line would for that data set. By graphing the line of best fit and looking for the choice that most resembles it, you can find the line with the greatest r^2-value.

For the TI-84:

To enter the data points, press [STAT] [1], and enter the x-values into L1 and the y-values into L2.

L1	L2	L3	3
1	1	
1	2	
2	1	
2	2	
3	2	
4	2	
4	3	

L3(1)=

To calculate the regression equation, press [STAT], go to the CALC tab, and press [4].

```
EDIT CALC TESTS
1:1-Var Stats
2:2-Var Stats
3:Med-Med
4:LinReg(ax+b)
5:QuadReg
6:CubicReg
7↓QuartReg
```

Make sure the Xlist is L1 and the Ylist is L2. If it is not already, you can enter L1 by pressing [2nd] [1] and L2 by pressing [2nd][2]. To plot the regression line, you can put Y1 in the Store RegEQ by pressing [VARS], going to the Y-VARS tab, and pressing [1] and [1] again. Then select Calculate.

```
LinReg(ax+b)
Xlist:L1
Ylist:L2
FreqList:
Store RegEQ:Y1
Calculate
```

The *a*- and *b*-values of the regression equation will be displayed.

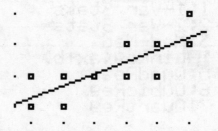

To see the scatterplot and the line of best fit together, press [2nd] [Y=] to get the STAT PLOTS menu. Press [1] and set to On. Then press [ZOOM] [9] [ZOOM] [5] to see the scatterplot. The line of best fit looks like that of line III in the question's scatterplot.

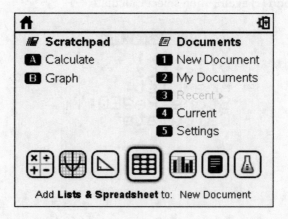

For the TI-Nspire:

From the home screen, select the Add Lists & Spreadsheet icon.

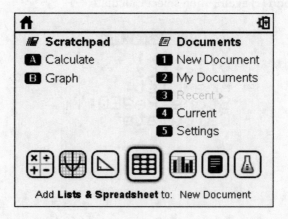

In the first row, label the A column "x" and the B column "y." Enter the *x*-values in cells A1 through A11. Enter the corresponding *y*-values in cells B1 through B11.

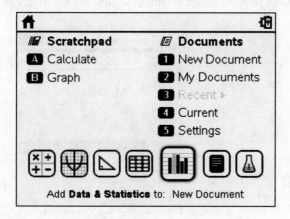

Go back to the home screen, and select the Add Data & Statistics icon.

To create the scatterplot correctly, click on the two "Click to add variable" buttons and set the horizontal axis to "x" and the vertical axis to "y." Your scatterplot should look like this:

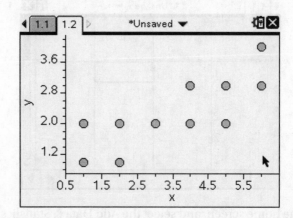

To see the line of best fit, select [menu], [4], and [6] for "Regression."

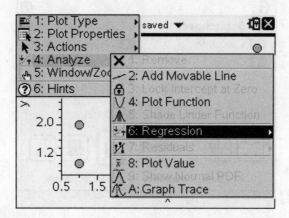

Then select [1] for "Show Linear (mx+b)" and press [enter].

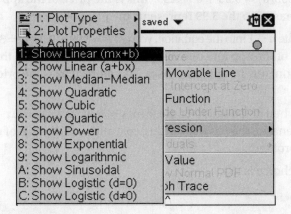

You will see the scatterplot along with the line of best fit. The line of best fit once again looks like that of line III in the question's scatterplot.

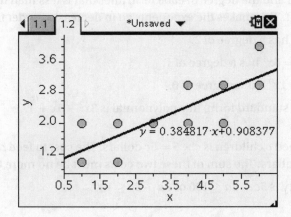

Of the four choices, line III matches most closely with the line of best fit. The line of best fit has the largest r^2-value.

The correct choice is **(3)**.

24. One way to answer this question is to substitute $n = 0$ into the equation and get $P = 3.99 + 0.99(0) = 3.99 + 0 = 3.99$. This is the price of the app without any in-app purchases, so the 3.99 represents the price of the app.

If you substitute $n = 1$ into the equation, you get $P = 3.99 + 0.99(1) = 3.99 + 0.99 = 4.98$. So, the price of the app and one in-app purchase is 0.99 more than the price of the app without any in-app purchases. Therefore, 0.99 is the price of one in-app purchase.

In general, when an equation of the form $y = b + mx$ or $y = mx + b$ is a model for a real-life scenario, the b is the starting cost and m is the cost of each additional item in the problem.

The correct choice is **(3)**.

Part II

25. A polynomial is in standard form when the term with the highest degree is the leftmost term and the degree of each term after that is less than the degree of the term to its left. This makes the exponents go in decreasing order from left to right.

The term $3x^2$ has a degree of 2.

The term $5x = 5x^1$ has a degree of 1.

The term $1 = 1x^0$ has a degree of 0.

Therefore, in standard form, this polynomial is $3x^2 + 5x + 1$.

26. The cost to feed c children is $c \times 5 = 5c$ dollars. The cost to feed a adults is $a \times 8 = 8a$ dollars. The sum of these two costs must be no more than \$100.

The inequality is $5c + 8a \leq 100$ or $8a + 5c \leq 100$.

27. To rationalize the denominator of a fraction when the denominator has a square root, multiply both the numerator and the denominator by the denominator and then simplify.

$$\frac{5}{\sqrt{3}} \cdot \frac{\sqrt{3}}{\sqrt{3}} = \frac{5\sqrt{3}}{\sqrt{3} \cdot \sqrt{3}} = \frac{5\sqrt{3}}{3}$$

$\frac{5\sqrt{3}}{3}$ is an expression with a rational denominator and is equivalent to the original expression.

28. The first step of this two-step algebra problem is to isolate the variable part by subtracting 10 from both sides of the equation.

$$\frac{2}{3}x + 10 = 6$$
$$-10 = -10$$
$$\frac{2}{3}x = -4$$

The next step is to isolate the x by multiplying both sides of the equation by the reciprocal of $\frac{2}{3}$, which is $\frac{3}{2}$.

$$\left(\frac{3}{2}\right)\frac{2}{3}x = \left(\frac{3}{2}\right)(-4)$$
$$1x = -6$$
$$x = -6$$

29. The first factor, $x^2 - 9$, can be factored with the difference of perfect squares factoring pattern: $a^2 - b^2 = (a - b)(a + b)$.

$$x^2 - 9 = x^2 - 3^2 = (x - 3)(x + 3)$$

So, the original equation can be written as the following:

$$(x - 3)(x + 3)(x - 2)(x + 5) = 0$$

If any of these factors is equal to zero, the equation will be true.

$x - 3 = 0$ means that $x = 3$ is a solution.

$x + 3 = 0$ means that $x = -3$ is a solution.

$x - 2 = 0$ means that $x = 2$ is a solution.

$x + 5 = 0$ means that $x = -5$ is a solution.

Therefore, all the solutions are $x = 3$, $x = -3$, $x = 2$, and $x = -5$.

30. To isolate the r, first divide both sides of the equation by P.

$$\frac{A}{P} = \frac{P(1 + r)^2}{P}$$
$$\frac{A}{P} = (1 + r)^2$$

Next, take the square root of both sides of the equation.

$$\sqrt{\frac{A}{P}} = \sqrt{(1 + r)^2}$$
$$\sqrt{\frac{A}{P}} = 1 + r$$

Finish by subtracting 1 from both sides of the equation.

$$\sqrt{\frac{A}{P}} = 1 + r$$

$$-1 = -1$$

$$\sqrt{\frac{A}{P}} - 1 = r$$

31. Since the total number of art students is 135, the number of students who take art class and prefer social studies must be $135 - 50 = 85$.

Since the total number of music students is 165, the number of students who take music class and prefer science must be $165 - 75 = 90$.

Add this information to the table.

	Art	Music	Total
Science	50	90	
Social Studies	85	75	
Total	135	165	300

The numbers in the total column can be calculated by adding each row (student who prefer science and students who prefer social studies).

	Art	Music	Total
Science	50	90	140
Social Studies	85	75	160
Total	135	165	300

Since $140 + 160 = 300$, these numbers don't lead to any contradictions.

To the *nearest tenth of a percent*, the probability that a student prefers social studies is equal to the number of students who prefer social studies divided by the total number of students, or $\frac{160}{300} = 53.3\%$.

If it is given that the student takes art class, the probability that the particular student prefers social studies is equal to the number of students who take art and prefer social studies divided by the total number of students who take art. T the *nearest tenth of a percent*, that is $\frac{85}{135} = 63.0\%$.

2. The fastest way to find the solution is to locate the intersection of the two lines. Since they intersect at the point (3, 6), the solution to this system of equations is $x = 3$, $y = 6$. You could also just write the ordered pair (3, 6).

Using algebra, one way to solve this system of equations is to use the elimination method. The elimination method is when you multiply both sides of one, or sometimes both, of the equations with the goal of making either the coefficients of the x-terms or of the y-terms opposites of each other. For this system, this can be accomplished by multiplying both sides of the first equation by 2.

$$2(-2x + 3y) = 2(12)$$
$$4x + 5y = 42$$
$$-4x + 6y = 24$$
$$4x + 5y = 42$$

Now you can add the two equations together and the x-terms will cancel out.

$$\frac{11y}{11} = \frac{66}{11}$$
$$y = 6$$

To solve for x, plug the y-value into either of the original equations.

$$4x + 5y = 42$$
$$4x + 5(6) = 42$$
$$4x + 30 = 42$$
$$-30 = -30$$
$$\frac{4x}{4} = \frac{12}{4}$$
$$x = 3$$

The solution is $x = 3$, $y = 6$, or (3, 6).

art III

3. The left side of an equation of the form $x^2 + bx = c$ can be turned into a perfect square trinomial by adding $\left(\frac{b}{2}\right)^2$ to both sides of the equation. For this question:

$$\left(\frac{b}{2}\right)^2 = \left(\frac{-24}{2}\right)^2 = (-12)^2 = 144$$

Add this to both sides of the equation.

$$x^2 - 24x + 144 = 5 + 144$$

The left side of this equation can now be factored since it is a perfect square trinomial.

$$(x - 12)^2 = 149$$

Take the square root of both sides of the equation.

$$\sqrt{(x - 12)^2} = \pm\sqrt{149}$$
$$x - 12 = \pm\sqrt{149}$$
$$x = \pm\sqrt{149} + 12 = 12 \pm\sqrt{149}$$

34. There are two ways to start this question. The longer but more straightforward way is to simplify the left side of the equation by multiplying and combining like terms.

$$(x + 5)^2 + 4(x + 5) - 12 = 0$$
$$(x + 5)(x + 5) + 4x + 20 - 12 = 0$$
$$x^2 + 5x + 5x + 25 + 4x + 20 - 12 = 0$$
$$x^2 + 14x + 33 = 0$$

Now that this is a basic type of quadratic equation, it can be solved by factoring, completing the square, or graphing.

Using factoring, this would become $(x + 3)(x + 11) = 0$. So either $x + 3 = 0$ or $x + 11 = 0$, resulting in $x = -3$ or $x = -11$. Therefore, the solution set is $\{-3, -11\}$.

There is a shortcut that you can use for this equation. Notice that $(x + 5)^2 + 4(x + 5) - 12 = 0$ has the same structure as the equation $u^2 + 4u - 12 = 0$, which would factor into $(u + 6)(u - 2) = 0$. So, the given equation can similarly factor into $((x + 5) + 6)((x + 5) - 2) = 0$, which simplifies to $(x + 11)(x + 3) = 0$. This leads to the same solution set $\{-3, -11\}$ as was determined using the longer method.

35. To find the median, first put the numbers in ascending (or descending) order.

$$9, 11, 12, 12, 13, 14, 35, 39, 41, 41, 65$$

There are 11 numbers, so the "middle" number of the list will be the 6th number since there are five numbers greater than the median and five numbers less than the median.

$$9, 11, 12, 12, 13, (14), 35, 39, 41, 41, 65$$

For this data set, the median is 14.

The first quartile is the median of all the numbers less than the median. For this set, the median of the numbers 9, 11, 12, 12, 13 is the 3rd number in the list, which is 12.

The third quartile is the median of all the numbers greater than the median. For this set, the median of the numbers 35, 39, 41, 41, 65 is the 3rd number in the list, which is 41.

When asked to create a box plot, first plot the five numbers: smallest value (minimum), first quartile, median, third quartile, largest value (maximum). Make small vertical line segments at each of the five points.

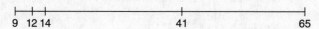

Then make a rectangle that has the segments for the first and third quartiles as two of its sides.

36. Start by creating a chart with function values between -4 and 2.

x	$f(x) = 2 \cdot 3^x$	$g(x) = 10$
-4	$f(-4) = 2 \cdot 3^{-4} = 2 \cdot \dfrac{1}{81} = \dfrac{2}{81} \approx 0.02$	10
-3	$f(-3) = 2 \cdot 3^{-3} = 2 \cdot \dfrac{1}{27} = \dfrac{2}{27} \approx 0.07$	10
-2	$f(-2) = 2 \cdot 3^{-2} = 2 \cdot \dfrac{1}{9} = \dfrac{2}{9} \approx 0.22$	10
-1	$f(-1) = 2 \cdot 3^{-1} = 2 \cdot \dfrac{1}{3} = \dfrac{2}{3} \approx 0.67$	10
0	$f(0) = 2 \cdot 3^0 = 2 \cdot 1 = 2$	10
1	$f(1) = 2 \cdot 3^1 = 2 \cdot 3 = 6$	10
2	$f(2) = 2 \cdot 3^9 = 2 \cdot 9 = 18$	10

When you graph the points, it looks like this.

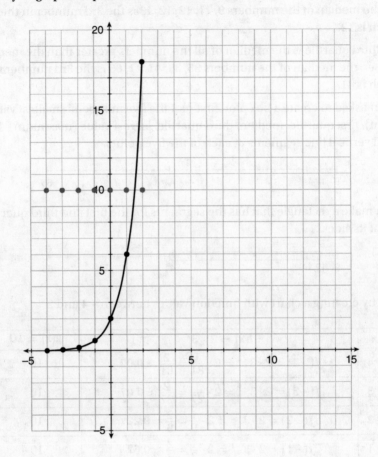

The point at which these two equations are equal to each other is the x-coordinate of the intersection of the two graphs. Looking at the graph shows that the intersection point occurs between $x = 1$ and $x = 2$.

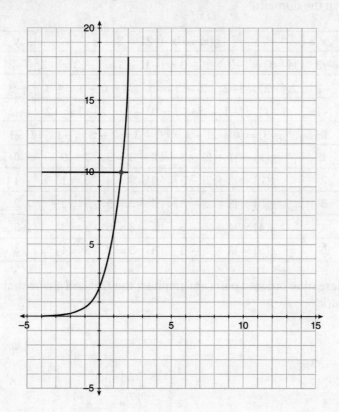

Part IV

37. The simplest way to create this graph is to make a chart with all the integer *x*-values in the domain.

x	g(x) = |x + 2| − 3	y
−3	$g(-3) = \|-3+2\| - 3 = \|-1\| - 3 = 1 - 3 = -2$	−2
−2	$g(-2) = \|-2+2\| - 3 = \|0\| - 3 = 0 - 3 = -3$	−3
−1	$g(-1) = \|-1+2\| - 3 = \|1\| - 3 = 1 - 3 = -2$	−2
0	$g(0) = \|0+2\| - 3 = \|2\| - 3 = 2 - 3 = -1$	−1
1	$g(1) = \|1+2\| - 3 = \|3\| - 3 = 3 - 3 = 0$	0
2	$g(2) = \|2+2\| - 3 = \|4\| - 3 = 4 - 3 = 1$	1
3	$g(3) = \|3+2\| - 3 = \|5\| - 3 = 5 - 3 = 2$	2
4	$g(4) = \|4+2\| - 3 = \|6\| - 3 = 6 - 3 = 3$	3
5	$g(5) = \|5+2\| - 3 = \|7\| - 3 = 7 - 3 = 4$	4

When these nine ordered pairs are graphed, they make a shape that resembles check mark.

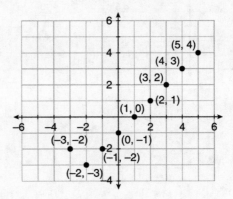

Fill in the rest of the solution set by drawing lines that pass through the points.

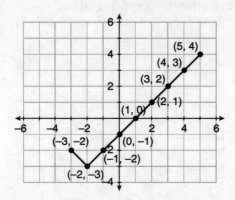

Another way this graph could have been created is to use function transformations. The graph of a simpler function could be $f(x) = |x|$. This function's graph looks like the letter V with the point of the V at $(0, 0)$, and the lines of the V have slopes of 1 and -1.

The function $g(x) = |x + 2| - 3$ can be thought of as $g(x) = f(x + 2) - 3$. The graph of the transformed function will be congruent to the original graph, but each point will be translated 2 units to the left and 3 units down.

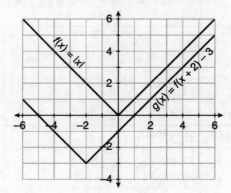

Since the domain of the function in this problem is $-3 \leq x \leq 5$, only the part of this new graph in that interval should be graphed.

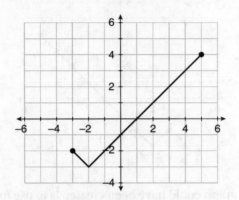

Topic	Question Numbers	Number of Points	Your Points	Your Percentage
1. Polynomials	17, 25	$2 + 2 = 4$		
2. Properties of Algebra	4, 28	$2 + 2 = 4$		
3. Functions	2, 5, 6, 9, 15, 18, 22	$2 + 2 + 2 + 2 + 2 + 2 + 2 = 14$		
4. Creating and Interpreting Equations	13, 24	$2 + 2 = 4$		
5. Inequalities	14, 26	$2 + 2 = 4$		
6. Sequences and Series	3, 21	$2 + 2 = 4$		
7. Systems of Equations	19, 32	$2 + 2 = 4$		
8. Quadratic Equations and Factoring	12, 20, 29, 30, 33, 34	$2 + 2 + 2 + 2 + 4 + 4 = 16$		
9. Regression	23	2		
10. Exponential Equations	10, 16	$2 + 2 = 4$		
11. Graphing	1, 36, 37	$2 + 4 + 6 = 12$		
12. Statistics	8, 11, 31, 35	$2 + 2 + 2 + 4 = 10$		
13. Number Properties	7, 27	$2 + 2 = 4$		
14. Unit Conversions				